MODIFICATION OF RADIOSENSITIVITY
OF BIOLOGICAL SYSTEMS

The following States are Members of the International Atomic Energy Agency:

AFGHANISTAN	HOLY SEE	PHILIPPINES
ALBANIA	HUNGARY	POLAND
ALGERIA	ICELAND	PORTUGAL
ARGENTINA	INDIA	QATAR
AUSTRALIA	INDONESIA	REPUBLIC OF SOUTH VIET-NAM
AUSTRIA	IRAN	ROMANIA
BANGLADESH	IRAQ	SAUDI ARABIA
BELGIUM	IRELAND	SENEGAL
BOLIVIA	ISRAEL	SIERRA LEONE
BRAZIL	ITALY	SINGAPORE
BULGARIA	IVORY COAST	SOUTH AFRICA
BURMA	JAMAICA	SPAIN
BYELORUSSIAN SOVIET	JAPAN	SRI LANKA
SOCIALIST REPUBLIC	JORDAN	SUDAN
CANADA	KENYA	SWEDEN
CHILE	KOREA, REPUBLIC OF	SWITZERLAND
COLOMBIA	KUWAIT	SYRIAN ARAB REPUBLIC
COSTA RICA	LEBANON	THAILAND
CUBA	LIBERIA	TUNISIA
CYPRUS	LIBYAN ARAB REPUBLIC	TURKEY
CZECHOSLOVAKIA	LIECHTENSTEIN	UGANDA
DEMOCRATIC KAMPUCHEA	LUXEMBOURG	UKRAINIAN SOVIET SOCIALIST
DEMOCRATIC PEOPLE'S	MADAGASCAR	REPUBLIC
REPUBLIC OF KOREA	MALAYSIA	UNION OF SOVIET SOCIALIST
DENMARK	MALI	REPUBLICS
DOMINICAN REPUBLIC	MAURITIUS	UNITED ARAB EMIRATES
ECUADOR	MEXICO	UNITED KINGDOM OF GREAT
EGYPT	MONACO	BRITAIN AND NORTHERN
EL SALVADOR	MONGOLIA	IRELAND
ETHIOPIA	MOROCCO	UNITED REPUBLIC OF
FINLAND	NETHERLANDS	CAMEROON
FRANCE	NEW ZEALAND	UNITED REPUBLIC OF
GABON	NIGER	TANZANIA
GERMAN DEMOCRATIC REPUBLIC	NIGERIA	UNITED STATES OF AMERICA
GERMANY, FEDERAL REPUBLIC OF	NORWAY	URUGUAY
GHANA	PAKISTAN	VENEZUELA
GREECE	PANAMA	YUGOSLAVIA
GUATEMALA	PARAGUAY	ZAIRE
HAITI	PERU	ZAMBIA

The Agency's Statute was approved on 23 October 1956 by the Conference on the Statute of the IAEA held at United Nations Headquarters, New York; it entered into force on 29 July 1957. The Headquarters of the Agency are situated in Vienna. Its principal objective is "to accelerate and enlarge the contribution of atomic energy to peace, health and prosperity throughout the world".

Printed by the IAEA in Austria
June 1976

PANEL PROCEEDINGS SERIES

MODIFICATION OF RADIOSENSITIVITY OF BIOLOGICAL SYSTEMS

PROCEEDINGS OF AN ADVISORY GROUP MEETING ON
MODIFICATION OF RADIOSENSITIVITY OF BIOLOGICAL SYSTEMS
ORGANIZED BY THE
INTERNATIONAL ATOMIC ENERGY AGENCY
AND HELD IN VIENNA, 8–11 DECEMBER 1975

RC 271
R3
A 33
1975

INTERNATIONAL ATOMIC ENERGY AGENCY
VIENNA, 1976

MODIFICATION OF RADIOSENSITIVITY
OF BIOLOGICAL SYSTEMS
IAEA, VIENNA, 1976
STI/PUB/446
ISBN 92–0–111176–2

FOREWORD

The treatment schedule of the radiotherapist has remained more or less unaltered during the last 30 years despite significant progress in the field of radiation biology. The radiotherapist, in fact, is overburdened with clinical work and rarely has enough time to think seriously about the new radiobiological concepts. He has adopted a pragmatic approach to his work and when he has had evidence of the efficacy of radiation treatment he has not delayed its application pending an agreement among radiobiologists on the theoretical principles behind it. When he has had a method of exposure that works, he has naturally shown little enthusiasm to change it.

Nevertheless, radiobiologists have continued to urge upon the clinician the need to try out their ideas, aimed at achieving better therapeutic results. They have suggested hyperbaric oxygen chambers and high LET radiations with a view to effectively destroying the hypoxic cells deeply embedded in the tumour. However, the giant accelerators and generators for producing high LET particles may prove too expensive for the developing countries. Under these circumstances, radioprotectors and hypoxic cell radiosensitizers may be useful alternatives. Whereas radiosensitizers would selectively enhance radiation damage to the cancerous cells, protectors can be used to minimize the harmful effects on the surrounding normal tissues.

The International Atomic Energy Agency has for some time been encouraging activities in this subject area. A panel of experts organized by the IAEA in collaboration with WHO discussed the radiosensitizing compounds in Stockholm in June 1973; the proceedings were published by the IAEA in 1974 under the title "Advances in chemical radiosensitization". The radioprotective compounds and their mechanisms of action had been discussed earlier at a panel held in Vienna in October 1968, the proceedings being published by the IAEA in 1969 under the title "Radiation damage and sulphydryl compounds".

New information has been accumulating which could be of particular relevance in the radiotherapy of cancer. New types of radiosensitizers and protectors have been discovered and the mechanisms of action have been better understood. Clinical trials initiated with some radiosensitizers have yielded encouraging results. It therefore seemed timely to discuss and evaluate these results with a view to providing guidelines for future research, and an Advisory Group on the Modification of Radiosensitivity of Biological Systems was called together by the IAEA in Vienna in December 1975. The papers presented as well as the conclusions and recommendations of the Group are included in the present Proceedings.

CONTENTS

A radiotherapist's view of radiosensitisers 1
  N.M. Bleehen
Radiation modifiers: an evaluation of recent research and clinical potential 11
  J.W. Harris
Radiation chemical basis for the role of glutathione in cellular radiation sensitivity 29
  M. Quintiliani, R. Badiello, M. Tamba, G. Gorin
Inhibition of DNA repair by chemical and biological agents 39
  H. Altmann, Helga Tuschl, E. Wawra, Ingrin Dolejs, W. Klein, A. Wottawa
Synergistic effect of radioprotective substances having different mechanisms of action 47
  L.B. Sztanyik, A. Sántha
Interference with endogenous radioprotectors as a method of radiosensitization 61
  S. Łukiewicz
MPG (2-mercaptopropionylglycine): a review on its protectove action against
  ionizing radiations 77
  T. Sugahara, P.N. Srivastava
Radioprotectors and radiotherapy of cancer 89
  J.R. Maisin, M. Lambiet-Collier, G. Mattelin
Nitroimidazoles as hypoxic cell sensitizers in vitro and in vivo 103
  G.E. Adams, J.F. Fowler
N_2O-mediated enhancement of radiation injury of *Escherichia coli* K-12 mutants
  in phosphate-buffered saline 119
  T. Brustad, E. Wold
Radiation sensitisation by membrane-specific drugs 131
  M.A. Shenoy, K.C. George, V.T. Srinivasan, B.B. Singh, K. Sundaram
Approaches to selective modification of radiation response 141
  E. Riklis, E. Ben-Hur
Radiation-dose dependence of sensitization by electron-affinic compounds 155
  L. Révész, B. Littbrand
Effect of Peptichemio on the survival of isolated mammalian cells 163
  O. Djordjević, L. Kostić
Interaction of Lucanthone and low radiation doses on mouse embryos in relation
  to LET and dose-rate 167
  Hedi Fritz-Niggli, C. Michel
Effect of Ro 07-0582 and radiation on a poorly reoxygenating mouse osteosarcoma 179
  L.M. van Putten, T. Smink
Clinical trials of Ro 07-0582 as a sensitizer of hypoxic cells 191
  I. Lenox-Smith, S. Dische
The problems of radiosensitization (short communication) 201
  E.F. Romantsev, A.V. Nikolskij
Changes in the ratio of activity of reparative and replicative enzymes of DNA
  synthesis as a basis for the search for radioprotective drugs (short communication) 205
  I.V. Filippovich, E.F. Romantsev

Conclusions and Recommendations 207
List of Participants 213

A RADIOTHERAPIST'S VIEW OF RADIOSENSITISERS

N.M. BLEEHEN
The Medical School,
Cambridge,
United Kingdom

Abstract

A RADIOTHERAPIST'S VIEW OF RADIOSENSITISERS.
Various approaches to the combination of drugs with radiation with the intent of producing a potentiating effect on tumour cells are discussed. The importance of consistent sensitisation of tumour tissue as opposed to normal tissue is emphasized. The possibilities of achieving a useful therapeutic gain factor for combined treatment with halogenated pyrimidines, electron-affinic hypoxic cell sensitisers, ICRF 159 and bleomycin are reviewed.

1. INTRODUCTION

Radiosensitisers have been the subject of numerous reviews {1-5} and of symposia {6-7} . The current panel will discuss a limited number of topics and this introductory paper is intended to define certain principles concerning the clinical use of such agents.

The treatment strategy of a radiation oncologist when faced with a cancer patient will depend on what he knows about the natural history of the particular type of tumour in terms of spread; the extent at the time of presentation; its known response to therapy and the general physical status of the patient. Thus diseases may conveniently be divided into two groups determined by their natural history. There are types where metastases occur early, such as with most lung cancers, bone sarcoma and many poorly differentiated tumours. In this group of patients the strategy for cure will not only depend on the ability to control local disease but also to influence the metastatic growth by adjuvant therapy using chemotherapeutic agents and, more speculatively, by immunotherapy or hyperthermia.

The other group of patients in which the disease remains localised for some time is the one in which radiotherapists and surgeons are most able to effect cures. There is still, however, around one third of patients with cancer who die with disease as a result of local treatment failure rather than because of distant spread. Methods of improving local control in these diseases by radiotherapeutic potentiation would therefore have a considerable impact in terms of total number of patients.

However, in spite of this rather arbitrary division of patients into two groups, one should not rule out completely the value of radiotherapeutic potentiation in diseases with a high metastic potential. Especially if it is also possible to treat the metastases successfully with chemotherapy. Thus effective prophylactic chemotherapy for osteosarcoma has been found which may well now influence the prognosis of this disease. Treatment with methotrexate and adriamycin and with other adjuvant agents, may eliminate occult metastases if given early enough {8-9} . Osteogenic sarcoma has usually been regarded as radioresistant and therefore surgical ablation is the primary treatment. However, radiotherapeutic cures are possible and have been reported in around 10% of cases {10}, this treatment usually being given because the tumour was not at a suitable site for amputation or surgery was refused. Most however fail to be controlled locally by radiotherapy. Goffinet and his colleagues {11} have reported 3 patients treated by radiotherapy following pulsed radiosensitisation with BUdR and associated with adjuvant chemotherapy. It may be that by using such techniques an improvement in local control will be obtained which, together with the adjuvant chemotherapy, will enable cures of osteosarcoma without limb ablation.

The groups of mechanisms by which drugs may produce an enhanced radiation effect on cells were defined in the conclusions of the previous panel {7}. In addition to these direct actions of drugs, one might also include other situations such as a drug which improves the abnormal tumour vasculature, as has been suggested for ICRF 159 {12-13}; or careful scheduling of a cytotoxic drug to produce shrinkage in a tumour associated with re-oxygenation, at which time radiotherapy should be more effective.

Many examples of drug and radiation combinations however only result in an adjunctive effect with an equal response of both tumour and normal tissue {5}. In the past too much attention has been paid to combinations of treatment which demonstrate radiosensitisation with scant regard for any adverse potentiation of the effect on normal tissues. The radiotherapist is not so worried about the technical problems of delivering enough radiation to a tumour to ablate it, unless this dose is accompanied by undue complications. Rads are inexpensive, patients are more precious.

We know that over a critically small dose range, relatively small increases in radiation dose may produce a considerable increase in cure rate {14-15}. A small potentiation of the radiation effect might then produce a disproportionate increase in cure rate. Unfortunately, the complication rate usually has a similar sigmoid shape. It is this relationship of the curve for tumour ablation and that for normal tissue complications which determines the clinical feasibility of radiotherapeutic cure.

The series of dose-response curves in Figure 1 is a theoretical representation of what might happen when radiosensitisation is attempted. Ideally one wishes to separate the response curves for tumour ablation and normal tissue complication rates so that there is no overlap, as in Fig. 1c. Merely to shift the curves equally to the left along the axis as in Fig. 1a will only save rads but not result in an improved therapeutic gain factor, as a given percentage of cures will still be associated with the same percentage of complications.

An adverse situation may occur if there is selective sensitisation of already sensitive tumour and of all the normal tissue cells, leaving resistant components of the tumour unsensitised. This might then result in curves like those seen in Fig. 1b.

It is therefore important to consider types and schedules of radiosensitisers and radioprotectors which are as fail safe as possible. If one is not going to do much good, at least one needs to be reasonably certain that not much harm will be done. Therefore in this context sensitisers of radioresistant components in tumours - such as hypoxic cell sensitisers, or protectors of euoxic cells are the most attractive current concepts.

The above considerations look at tumour ablation and normal tissue complications as abstract phenomena. Radiation therapists are concerned with the attempted cure of a volume of tumour contained within a viable host. The term complication may then cloak a variety of clinically acceptable or unacceptable situations. The volume of tissue treated must also be considered. Major complications in a small volume of tissue may not be as troublesome as less serious complications throughout a larger volume.

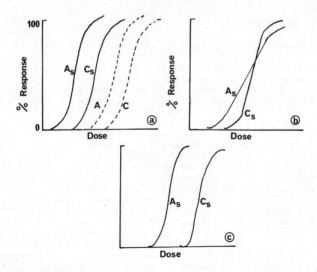

FIG.1. Theoretical curves for tumour ablation and normal tissue complication rates with increasing radiation dose.

    A = tumour ablation

    C = normal tissue complication

    A_C, C_S = rates with sensitiser

(a) Equal effect of sensitiser on tumour and normal tissue.
(b) Variable effect of sensitiser on tumour.
(c) Marked effect of sensitiser on tumour only.

    The clinician might therefore find some sensitisation of normal tissue acceptable in a small volume with attendant complications, if it also results in tumour ablation. However in the larger tumour, when a larger dose to control 90% of the tumours (TCD_{90}) might be expected {16}, normal tissue sensitisation would be wholly unacceptable because of the volume of damaged normal tissue.

2. HALOGENATED PYRIMIDINES

    The group of drugs most extensively investigated in man as chemotherapeutic and radiosensitising agents are the halogenated pyrimidines and their nucleosides { 1, 2, 17 }.
    None of the major clinical studies provide convincing clinical evidence of potentiation when 5-Flurouracil is used in conjunction with radiotherapy. Other halogenated pyrimidines may act as true radiosensitisers and there have been clinical trials utilizing 5-BUdR or 5-IUdR at several different tumour sites {1, 5}. Results have been variable, and the results so far with halogenated pyrimidines do appear therefore to be disappointing. Perhaps, with careful selection of regional disease accessible to arterial perfusion and suitable scheduling of the sensitisers as proposed by Brown and his colleagues {18} this technique may be of some limited value.

3. HYPOXIC CELL SENSITISERS

I should now like to turn to another topic which to me is perhaps the most exciting one for clinical radiotherapy at this time. Certain electric affinic substances may selectively sensitise hypoxic cells without any sensitisation of the normal cells {4}.

Two most promising drugs that have been particularly investigated are Metronidazole and a Roche product which is known by the code of Ro-07-0582. Both these agents are relatively non-toxic to cells in vitro and can be given in high doses to animals and man {19, 20}. Metronidazole given in large doses by mouth may approach the same peak serum concentration as that required to achieve sensitisation in mouse tumours {21}. The drug is relatively nauseating but no untoward toxicity in man has been demonstrated, from large single doses. Successful phase I studies for repeated doses of Metronidazole have also recently been reported {22}. Radiotherapeutic studies have been commenced using it but no definite conclusions have yet been reported, although no evidence of adverse normal tissue effect has been observed.

Considerable interest now is centred around one of the 2-nitroimidazoles. Ro-07-0582 has been shown, in a variety of systems, to be even more effective as a sensitiser of hypoxic cells than Flagyl and like it, also to be relatively metabolically stable {19}. We {23} have found an enhancement ratio of 2.2 using the EMT6 mouse mammary tumour treated in vivo with a single fraction of radiation (Fig. 2). The animals were given 1mg/gm of Ro-07-0582 by the intraperitoneal route, 30 minutes before irradiation. Of course, if one can achieve adequate reoxygenation employing suitably fractionated radiation, this sensitisation then becomes less significant.

It is too early to bring this drug into routine clinical practice but the animal {19} and preliminary clinical results to be reported later {20} do lead one to hope that this might prove to be a useful agent.

4. ICRF 159

A chemotherapeutic agent of recent experimental interest is a bisdioxopiperazine, ICRF 159, {24} synthesised in the laboratories of the Imperial Cancer Research Fund, London. This bisdioxopiperazine has been reported to be a potent inhibitor of DNA synthesis and also to block progression through the cell cycle. Its effect is probably confined to one part of the cell cycle - the transition between G_2 and M.

Its usefulness in man for cancer chemotherapy used as a single agent has been disappointing because of its toxicity to normal tissues. However, its interest for radiation therapy lies in other properties of its action. It has been shown to induce changes in the blood vessels of some tumours, resulting in normalisation of the previously abnormal tumour vasculature {25}. This may then, as in experiments with the Lewis lung tumour, be associated with a reduction in the number of metastases. The drug has been shown to have some potentiation of the effect of radiation on the S180 tumour in rats {12}. However in vitro treatment of HeLa cells by ICRF 159 immediately followed by X-radiation failed to demonstrate any such radiosensitisation {26}. It has therefore been concluded that some at least of the in vivo radiosensitisation with the S180 tumour might have been associated with a normalised blood supply and improved oxygenation {12}.

Other workers {13} have similarly studied the Walker 256 carcinosarcoma and found an increased effect of the combination of ICRF 159 and radiation, but could not exclude an additive effect.

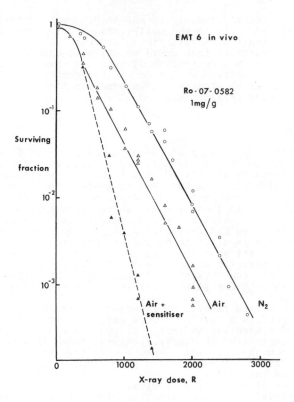

FIG. 2. Radiation dose response curve for EMT6 mouse tumour irradiated in vivo and assayed in vitro for surviving fraction.

○ *animals killed and irradiated in nitrogen*

△ *animals breathing air*

▲ *animals breathing air and given 1 mg/g of Ro-07-0582 intraperitoneally 30 min before irradiation.*

They quote data of Baungärtl and colleagues that the improved vascularisation following ICRF 159 treatment does lead to an increase in tissue pO_2, in this tumour.

Because of the possible clinical interest of this form of radiosensitisation we have been carrying out similar work and have been able to demonstrate a modest change in Do when exponentially growing EMT6 cells are exposed to the drug in vitro for 24 hours before X-radiation {27}. This effect is not seen with shorter drug exposures of 1 hour which probably explains Dawson's inability to demonstrate sensitisation in vitro. It does indicate that some radiosensitisation may occur without having to invoke improved oxygenation due to a change in tumour blood supply. Flow cytoflurographic evidence suggests that after 24 hours exposure to ICRF 159, there is a considerable build up of cells in G_2. This may be one explanation for the change in the Do at that time, although current work in progress leads us to doubt this as the sole explanation.

Several clinical studies are now in progress to test this possible synergism. Ryall and colleagues {28} reported on a series of 22 patients with soft tissue and bone sarcomas. They claimed better responses than could normally be expected. Severe toxicity resulting in interruption of treatment was not seen, although the skin reactions were greater than expected. The results of controlled clinical studies now in progress should show if this clinical impression is substantiated.

5. BLEOMYCIN

Yet another approach that has been investigated is the possibility that bleomycin might act as a radiosensitising agent. Bleomycin is known to inhibit DNA synthesis and to produce DNA strand breaks {29-30}. It has been suggested that it might have a synergistic effect on that of X-radiation { 5 }. This could be of particular value in those tumours where bleomycin is known to be useful in therapy, such as well differentiated squamous carcinoma in the head and neck region, or in those tumours where bleomycin has been reported to be selectively concentrated. This selective concentration of the drug could then provide an amplification factor for any possible radiosensitisation. Thus an increase in the concentration of radioactive bleomycin in an experimental mouse brain glioma over that in normal brain has been reported {31-32}.

There have been reports of responses of primary brain tumours to bleomycin used as a carcinolytic agent. In view of this, and the possibility of concentration in the gliomas of man, we studied the bleomycin concentration in biopsy specimens taken from patients with glioblastoma multiforme. When possible we looked at glioma tissue and a sample of normal brain which the surgeon removed during the approach to the biopsy excision. In all 5 patients, within the limitations of the microbiological assay method { 33}, the glioma showed an increase in concentration of bleomycin of between 2-12 with a mean ratio of 5 times that of adjacent normal brain.

Unfortunately the investigations for possible radiosensitisation by bleomycin do not appear to give a clear answer either when using in vitro cell culture systems or following treatment of tumours in vivo.

We have studied the effect of the combination of bleomycin and X or gamma radiation in bacterial and two mammalian cell lines {34}. There was a marked sensitisation effect when using the radiation resistant E. coli B/r. However, this effect was only seen when the cells were exposed to the drug after the radiation and was not seen when the exposure to bleomycin was before or during the radiation treatment. This mutant has a high capacity for removing radiation induced DNA single strand breaks and it may be that bleomycin acts by reducing the repair capacity of this strain. No such potentiation effect was seen in two other bacterial strains used, E. coli B/s and Micrococcus radiodurans. Likewise no sensitisation was seen by us with two mammalian cell lines in vitro, HeLa and the EMT6 mouse mammary tumour.

Bienkowska and her colleagues {35} were unable to find any potentiating effect of bleomycin on X-radiation using HeLa cells in vitro. However, Matsuzawa and his colleagues {36} using a mouse mammary carcinoma line showed what appears to be a reduction in the shoulder width, and possibly in the slope, after pretreatment of the cells for 1 hour with a dose of bleomycin which causes a small amount of cell killing (around 10%). In view of the small changes seen their conclusions must be regarded with some caution.

The results of in vivo experiments are also contradictory. Jørgensen {37} presented experimental evidence that a squamous carcinoma in mice may be controlled more successfully when bleomycin is added to the radiotherapy. Radiation alone produced a 70% reduction of tumour growth in a 3/52 period after treatment. The combination treatment with bleomycin resulted in over 95% regression. However, no indication was given of any local increase in reaction in the skin, and there is no evidence therefore that any useful gain factor was obtained. In contrast, Sakamoto and Sakka {38} were unable to demonstrate any sensitising effect on the radiation response of a murine squamous carcinoma treated in vivo and assayed by the TD_{50} dilution assay method.

In spite of this experimental evidence there are numerous reports of the continued use of radiotherapy together with bleomycin in treatment, principally for squamous carcinoma in the head and neck region. There are enthusiastic results claimed but they are difficult to evaluate in the absence of suitable controls. The Medical Research Council in the United Kingdom is now carrying out a controlled trial to investigate this problem.

CONCLUSIONS

This paper summarises a few of the possible approaches to radiosensitisation that are now being investigated. These and others are the subject of more detailed analysis in subsequent papers presented at this panel. The great problem for the radiotherapist with most studies previously reported has been the lack of tumour specificity. There are experimental and early clinical indications that this goal may now be in sight.

REFERENCES

{1} DOGGETT,R.L.S., BAGSHAW, M.A., KAPLAN, H.S., "Combined therapy using chemotherapeutic agents and radiotherapy" In: Deeley, T.J., Wood, C.A.P., ed Modern Trends in Radiotherapy 1 (1967) 107 Butterworths, London.
{2} VERMUND, H., GOLLIN, F.F., Mechanisms of action of radiotherapy and chemotherapeutic adjuvants, Cancer, 21 (1968) 58.
{3} BERRY, R.J., "Radiotherapy plus chemotherapy - have we gained anything by combining them in the treatment of human cancer?" Frontiers of Radiation Therapy and Oncology 4 (1969) 1.
{4} ADAMS, G.E., Chemical radiosensitisation of hypoxic cells, Brit. Med. Bull. 29 (1973) 48.
{5} BLEEHEN, N.M., Combination therapy with drugs and radiation, Brit. Med. Bull. 29 (1973) 54.
{6} MOROSON, H., QUINTILIANI, M. ed. Radiation protection and sensitisation 1969, Taylor and Francis, London.
{7} International Atomic Energy Agency Panel. Advances in chemical radiosensitisation 1974, I.A.E.A. Vienna.
{8} JAFFE, N., FREI, E., TRAGGIS, D., BISHOP, Y., Adjuvant methotrexate and alrovorum factor treatment of osteogenic sarcoma, N. Engl. J. Med. 291 (1974) 994.
{9} CORTES, E.P., HOLLAND, J.F., WANG, J.J., SINKS, L.F., BLOM, J., SENN, H., BANK, A., GLIDEWELL, O., Amputation and adriamycin in primary osteosarcoma, New Eng. J. Med. 291 (1974) 998.
{10} FRIEDMAN, M.A., CARTER, S.K., The therapy of osteogenic sarcoma current status and thoughts for the future, J. Surg. Oncol. 4 (1972) 482.
{11} GOFFINET, D.R., KAPLAN, H.S., DONALDSON, S.S., BAGSHAW, M.A., WILBUR, J.R., Combined radiosensitiser infusion and irradiation of osteogenic sarcomas, Radiology 117 (1975) 211.
{12} HELLMANN, K., MURKIN, G.E., Synergism of ICRF 159 and radiotherapy in treatment of experimental tumours, Cancer 34 (1974) 1033.

{13} NORPOTH, K., SCHAPHAUS, A., ZIEGLER, H., WITTING, V., Combined treatment of the Alker tumour with radiotherapy and ICRF 159, Z. Krebsforsch 82 (1974) 329.
{14} SHUKOVSKY, L.J., Dose, time, volume relationships in squamous cell carcinoma of the supraglottic larynx, Amer. J. Roengentol. 108 (1970) 27.
{15} MORRISON, R., The results of treatment of cancer of the bladder - a clinical contribution to radiobiology, Clin. Radiol. 26 (1975) 67
{16} FLETCHER, G.H., Clinical dose-response curves of human malignant epithelial tumours, Brit. J. Radiol. 46 (1973) 1.
{17} SZYBALSKI, W., X-ray sensitisation by halopyrimidines, Cancer Chemother. Rep. 58 (1974) 539.
{18} BROWN, J.M., GOFFINET, D.R., CLEAVER, J.E., KALLMANN, R.F., Preferential radiosensitisation of mouse sarcoma relative to normal skin by chronic intra-arterial infusion of halogenated pyrimidine analogues, J. Natl. Cancer Inst. 47 (1971) 75.
{19} ADAMS, G.E., FOWLER, J.F., Nitroimidoyles as hypoxic cell sensitisers in vitro and in vivo, I.A.E.A. Panel on Radiosensitisers and Protectors, Vienna 1975.
{20} LENOX-SMITH, I., DISCHE, S., Clinical trials of Ro-07-0582 as a sensitiser of hypoxic cells, I.A.E.A. panel on Radiosensitisers and protectors, Vienna 1975.
{21} DEUTSCH, G., FOSTER, J.L., McFADZEAN, J.A., PARNELL, M., Human studies with "high dose" metronidazole: a non-toxic radiosensitiser of hypoxic cells, Brit. J. Cancer 31 (1975) 75.
{22} URTASUN, R.C., CHAPMAN, J.D., BAND, P., RABIN, H.R., FRYER, C. G., STURMURIND, J., Phase I study of high dose metronidazole: a specific in vivo and in vitro radiosensitiser of hypoxic cells, Radiology 117 (1975) 129.
{23} BLEEHEN, N.M., HAR-KEDAR, I., WATSON., J.V., - unpublished data.
{24} CREIGHTON, A.M., HELLMANN, K., WHITECROSS, S., Antitumour activity in a series of bisDiketopiperazines, Nature 222 (1969) 384.
{25} LeSERVE, A.W., HELLMANN, K., Metastases and the normalisation of tumour blood vessels by ICRF 159: A new type of drug action, Brit. Med. J. 2 (1972) 597.
{26} DAWSON, K.B., - unpublished data quoted in {13}.
{27} TAYLOR, I., BLEEHEN, N.M., - unpublished data.
{28} RYALL, R.D.H., HANHAM, I.W.F., NEWTON, K.A., HELLMANN, K., BRINKLEY, D.M., HJERTAAS, O.K., Combined treatment of soft tissue and osteosarcomas by radiation and ICRF 159, Cancer 34 (1974) 1040.
{29} SUZUKI, H., NAGAI, K., YAMAKI, H., On the mechanism of action of bleomycin scission of DNA strands in vitro and in vivo, J. Antibiotics 22 (1969) 446.
{30} TERASIMA, T., YAKUKAWA, M., UMEZAWA, H., Breaks and rejoining of DNA in cultured mammalian cells treated with bleomycin, Gann 61 (1970) 513.
{31} KANNO, T., Experimental studies on nonoperative treatment of malignant brain tumours - Distribution of anticancer drugs in the organs of mice with experimental brain tumours, Cancer Chemother. Abstr. 12 (1971) 464.
{32} HAYAKAWA, T., USHIO, Y., MOGAMI, H., HORIBATA, K., The uptake, distribution and anti-tumour activity of bleomycin in gliomas of mice, Europ. J. Cancer 10 (1974) 137.
{33} PITTILLO, R.F., WOOLLEY, C., RICE, L.S., Bleomycin, an antitumour antibiotic - Improved microbiological assay and tissue distribution studies in normal mice, Applied Microbiol. 22 (1971) 564.
{34} BLEEHEN, N.M., GILLIES, N.E., TWENTYMAN, P.R., The effect of bleomycin and radiation on bacteria and mammalian cells in culture, Brit. J. Radiol. 47 (1974) 346.

{35} BIENKOWSKA, Z.M., DAWSON, K.B., PEACOCK, J.H., Action of actinomycin D, bleomycin and X rays on HeLa cells, Brit. J. Radiol. 46 (1973) 619.
{36} MATZUZAWA, T., ONOZAWA, M., MORITA, K., KAKEHI, M., Radiosensitisation of bleomycin on lethal effect of mouse cancer cells in vitro, Strahlentherapie 144 (1972) 614.
{37} JØRGENSEN, S.J., Time-dose relationships in combined bleomycin treatment and radiotherapy, Europ. J. Cancer 8 (1972) 531.
{38} SAKAMOTO, K., SAKKA, M., The effect of bleomycin and its combined effect with radiation on murine squamous carcinoma treated in vivo, Brit. J. Cancer 30 (1974) 463.

RADIATION MODIFIERS
*An evaluation of recent research
and clinical potential*

J.W. HARRIS
University of California,
San Francisco, California,
United States of America

Abstract

RADIATION MODIFIERS: AN EVALUATION OF RECENT RESEARCH AND CLINICAL POTENTIAL.
    Although radiation-modifying agents have made significant contributions to our understanding of basic radiobiological mechanisms, their impact on clinical radiotherapy has been limited. Recent developments in protection and sensitization, however, have raised new hopes that some of these agents may soon find a place in therapy and have enticed a number of researchers and clinicians to seriously evaluate this possibility. The paper reviews these recent developments, with emphasis on work in the United States, and makes recommendations for experimental areas and approaches which appear to have considerable promise — in particular, combinations of sensitizing and protective agents, of sensitizers that act by independent mechanisms, and of modifying agents and other treatment modalities. It is concluded that intensification of basic efforts, in conjunction with well-designed and carefully evaluated clinical trials, could lead, within a relatively short time, to a definitive evaluation of the importance of anoxic cells in human radiotherapy and to a significant clinical role for radiation-modifying compounds.

I. INTRODUCTION

    For more than twenty-five years, scientists in many countries have studied chemical compounds that modify the effects of ionizing radiation on biological systems. Although these studies have contributed significantly to our basic knowledge about radiation effects, they have not yet had an impact on cancer radiotherapy and, until fairly recently, the majority opinion was that they never would. For whatever reasons, interest in the area flagged and many workers turned to less frustrating interests.
    In recent years, this trend has been dramatically reversed and the chemical modification is now enjoying something of a renaissance. Subjectively, all of us who referee for scientific journals are aware that the number of manuscripts in radiation modification grows larger each month. In more objective terms, the rate of publication in the field provides a good indicator of this trend: Figure 1, for example, documents the intense activity generated through the 1960s and the slump that followed. However, when one examines the year-by-year publication of papers in only <u>biological</u> areas, it is apparent that both protectors and sensitizers are enjoying a resurgence of interest (see Figure 1). With this renewed interest has come renewed hope for eventual clinical applications.
    This meeting provides a unique opportunity to assess the present "state of the art" and to make recommendations for future directions. It is particularly appropriate that we have been convened by the IAEA, because of the great potential contribution to this area of research by scientists and clinicians in less developed nations, who may not have ready access to facilities such as neutron generators.

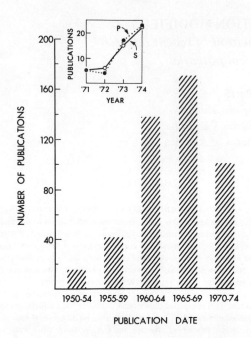

FIG.1. *Publication of journal articles on radiation protection, 1950–1974. The number of papers published on radiation chemistry and biology of chemical modifiers is compiled from references cited by Foye* [1] *and by Klayman and Copeland* [2]. *The insert shows publications on the effects of protective agents (P) or sensitizers (S) on mammals or mammalian cells and tissues during 1971–1974 (data taken from a MEDLINE computer search and the author's personal files).*

In keeping with this broader interpretation of our purpose here, I will devote my paper to a review of recent progress in selected areas of both protection and modification, with particular emphasis on two areas which other speakers may not touch on in detail: chemical protection with thiophosphates and chemical sensitization with agents other than the so-called "electron-affinic" compounds, which I am sure will be well reviewed by others. As the only speaker from the Americas, I feel obliged to focus particularly on developments in that part of the world -- but I hope you will view this as appropriate rather than exclusive.

II. RECENT DEVELOPMENTS IN CHEMICAL PROTECTION

Much of the resurgence in interest in radioprotection, in the United States at least, can be traced to the discovery by Yuhas and Storer [3] that the thiophosphate compound S-2-(3-aminopropylamino) ethylphosphorothioic acid (WR-2721) protected irradiated tissues in a tumor-bearing mouse but did not protect the tumor itself. Other lines of research have also been pursued -- for example, protection of specific organs by infusion of pharmacologic agents such as vasopressin [4] and catecholamines [5,6] -- but thiophosphates have clearly been the major theme. The most prominent of these compounds are illustrated in Table I.

TABLE I. CHEMICAL STRUCTURE OF CYSTEAMINE-RELATED THIOPHOSPHATE RADIOPROTECTIVE COMPOUNDS

Designation	Chemical name	Structure
MEA	β-mercaptoethylamine (cysteamine)	$H_2NCH_2CH_2SH$
WR-2529	3-(2-mercaptoethylamino) propionamide ρ-toluenesulfonate	$HSCH_2CH_2NH\ CH_2CH_2CNH_2 \cdot CH_3C_6H_4SO_3H$ (with C=O on the middle C)
WR-2721	S-2-(3-aminopropylamino)ethyl phosphorothioic acid hydrate	$H_2NCH_2CH_2CH_2\ NHCH_2CH_2S\ PO_3H_2 \cdot XH_2O$
WR-2823	S-2-(5-aminopentylamino) ethyl phosphorothioic acid monohydrate	$H_2N(CH_2)_5\ NHCH_2CH_2S\ PO_3H_2 \cdot H_2O$
WR-638	Sodium hydrogen S-(2-aminoethyl) phosphorothioate	$H_2NCH_2CH_2S\ \overset{O}{\underset{\|}{P}}\!\!-\!(OH)(ONa)$

TABLE II. PROTECTION OF MOUSE TISSUES BY WR-2721 INJECTED 15-30 min BEFORE X-RADIATION

Endpoint	WR-2721 dose (mg/kg)	D.M.F.	Reference
$LD_{50/30}$	500	2.6-2.7	3
Skin ulceration (30 days)	"	2.4	"
$LD_{50/0}$ (CNS)	250,500	1.0, 0.5	7
$LD_{50/7}$ (GI)	"	1.6, 1.8	"
$LD_{50/30}$ (marrow)	"	2.3, 2.7	"
Development of immune cells	500	3.4	8
$LD_{50/120}$ (lung death)	"	1.7	9
Hair loss (60 days)	"	2.1	"
Marrow skin killing (aerobic)	600	3.0	10
Marrow skin killing (hypoxic)	"	1.6	"
$LD_{50/30}$	500-600	2.2	11
Esophageal lethality	"	1.4	"
Pulmonary lethality	"	1.2	"
Renal lethality	"	1.5	"
$LD_{50/6}$ (GI)	500	1.64	12
Acute skin reaction	400	2.0	13,14
Late tissue changes	"	1.5	"

Yuhas and Storer's observation [3] that WR-2721 protected mice against radiation skin damage (DMF=2.4) and 30 day-lethality (DMF=2.7) but did not modify the radiosensitivity of a transplanted mammary tumor stimulated considerable interest. These same authors demonstrated that WR-2721 and related compounds were less effective against gastrointestinal death than against marrow death and were completely ineffective in protecting against central

TABLE III. PROTECTION OF ANIMAL TUMORS BY WR-2721 INJECTED 15-30 min BEFORE X-IRRADIATION

Endpoint	WR-2721 dose (mg/kg)	D.M.F.	References
% transplantability of mammary carcinoma	500	1.15	3
Lung tumor volume	400	55% of control	9
EMT-6 tumor cell killing (aerobic)	500	1.5-2.0	15
EMT-6 tumor cell killing (acutely hypoxic)	"	1.2-1.5	"
EMT-6 tumor cell killing (chronically hypoxic)	"	~1.0	"
P388 leukemia cell killing	0.005 mmoles	1.6	10
P388 leukemia (mean survival time)	500-600	2.2	11
EMT-6 tumor (cure)	"	1.3	"
Control of KHT sarcoma	400	1.2	14

nervous system death [7]. Moreover, they found significant strain differences in the degree of protection by WR-2721 [16] and suggested that this might relate to dephosphorylation of the compound by serum enzymes.

These early studies were soon followed by others, characterizing the effects of WR-2721 on various normal tissues (Table II) and tumors (Table III). Our own studies showed that the compound protected marrow colony-forming units better than cysteamine did [10], the DMFs being 3.0 and approximately 1.7, respectively. Moreover, WR-2721 protected hypoxic marrow cells only slightly and, unlike cysteamine, entered cells by passive diffusion [10]. Utley et al. [15] confirmed, in the mouse EMT-6 tumor, that WR-2721 was a poor protector for hypoxic cells (Table III). Taken together, these results indicated that the differential protection of normal and malignant tissues described by Yuhas and Storer [3] could probably be explained by the relative inability of WR-2721 to penetrate into poorly vascularized tumors in sufficient quantities to result in a radiobiologically significant intracellular concentration and by its failure to protect hypoxic (and particularly chronically hypoxic) cells. Extensive biochemical comparisons of normal and malignant tissues disclosed no inherent differences in the ability of tumor vs non-tumor tissue to dephosphorylate WR-2721 and related compounds to the radioprotective free-SH form [10].

A number of workers have recently extended the comparison between normal tissue and malignant tissue effects of WR-2721 to other systems (Table II and III). It became clear early on that the compound did protect tumor cells in the ascites form [10,11] and would not, therefore, be useful in the treatment of hematologic neoplasms. With this single exception, however, the original observation of poor tumor protection has generally been confirmed (Table III).

Phillips, Kane and Utley [11] obtained a DMF of 1.3 for cure of solid EMT-6 carcinomas in mice, although they also found a disturbingly wide range of DMFs for protection of various normal tissues (1.2 for lung death to 3 for bone marrow CFUs). (Their low DMF for lung contrasts with a considerably higher one obtained by Yuhas [9], but he used specific-pathogen-free mice.) Yuhas [9] showed that the selective protection of WR-2721 for normal tissue applied also to a urethan-induced lung tumor system (Table III), (although the tumors were protected when the interval between drug injection and irradiation was extended to 1-2 hours) and Lowy and Baker [14] obtained equally encouraging results with the KHT tumor system.

Although these single-dose studies are highly encouraging, they are not particularly germane to the clinical situation, where radiation is routinely delivered in multiple fractions over a period of many days. Very recently, a few workers have begun to examine the effects of WR-2721 in multiple fraction experiments, with promising results. Utley, Phillips and Kane [21] administered WR-2721 to mice 30 minutes before each of ten irradiations (in 12 days) and obtained DMFs of 1.3-1.5 for skin damage (compared to 1.5-1.7 for single dose exposure) and 1.7 for intestinal damage (compared to 1.6 for the single dose). They found it necessary, however, to decrease the amount of protective agent they administered before each fraction in order to avoid acute toxicity. This problem might also be circumvented by increasing the time between fractions. Other work in progress in this area includes that of D. Biery (School of Veterinary Medicine, University of Pennsylvania). Biery is irradiating Beagle dogs to a total dose of 4000 rads (4x1000) through a 3x3 cm portal of the right lateral maxilla, either with or without intravenous administration of 75 mg/kg WR-2721 before irradiation. His results, which include animals treated for spontaneous oral fibrosarcomas, squamous cell carcinomas and malignant melanomas, indicate significant protection of skin and oral mucose by the protective agent [personal communication]. This protection permits delivery of increased radiation doses to the tumors; there is, to date, no evidence that WR-2721 alters the radiation response of the neoplasms treated.

Another potential application of WR-2721 is in protecting normal tissues in patients treated with high LET radiations. Only a few experiments with high LET have been reported, but they are encouraging. Sigdestad and co-workers compared the protective efficacy of WR-2721 for the gastrointestinal tract of mice irradiated with 4-MeV Xrays or with fission neutrons. Their data agree with that of other workers in indicating relatively low protection for X-irradiated gut (DMF=1.64) and they obtain a similar DMF (1.6) for neutron-irradiated gut. Recent studies by Baker and Leith [23] show that high LET damage to skin (in this case from helium ions) can be moderated by topical application of thiols. In view of the considerable tissue reactions sometimes associated with high LET radiation therapy, these results should encourage further examination of chemical protectors against high LET radiation damage.

Fairly extensive studies have been conducted to determine the pharmacologic effects and distribution of WR-2721 and other thiophosphates, but only a few of these have yet reached the published literature. The compound causes profound vasodilation in mice, raising the possibility that reduction of peripheral oxygen tension may contribute to its superiority as a radioprotector [17], but has little acute cardiovascular effect in cats and dogs [18]. Dogs treated with WR-2721 do develop a gradual increase in blood pressure which is not due to norepinephine-induced vasoconstriction, however, and there is clear evidence that the compound is an effective ganglionic blocking agent [18]. The significance of these observations for potential clinical use of WR-2721 is not clear.

The distribution of ^{35}S-labeled WR-2721 in normal and malignant tissues of mice and rats has recently been studied [19]. Taken together with earlier work on related agents [20], these studies indicate that the compound does not penetrate into the brain but easily reaches most other tissues, including intestines and lung. This finding leaves the failure of WR-2721 to protect the latter two tissues [e.g., 11] unexplained. Washburn et al. [19] examined the uptake of labeled WR-2721 by four different types of rat and mouse tumors and found that three of them contained less compound than liver, lung, kidney, small intestine or spleen. No explanation was offered for the failure of the fourth tumor (the Morris 7777 hepatoma) to show decreased uptake and the relative radioprotectability of the four tumors was not reported.

The mechanism(s) by which WR-2721 protects at the cellular/molecular level has not been studied in detail, partly because the compound does not protect cells in vitro [unpublished data]. What data is available suggests that WR-2721 is dephosphorylated enzymatically as soon as it penetrates the cell membrane and that the free-SH compound thus formed protects by the same mechanisms as do cysteamine and other thiols [e.g., 10,24,25]. The limited number of molecular-level studies reported to date [e.g., 26] give no reason to suspect that this conclusion is incorrect. The identity of these mechanisms, a matter of long-standing and intense dispute, is outside the scope of this review, though I do wish to refer the reader to recent publications from our laboratory and elsewhere [see ref. 27] that cast considerable doubt on the role traditionally assigned to mixed disulfide formation.

Finally, recent work on an alternate approach to radiation protection should be mentioned. It has been realized, for some time, that topical application of radioprotective thiols can reduce skin and mucosal radiation reactions by DMFs of 1.2-1.7 [13,28-30]. WR-2721 has no effect when applied topically [13]. This work has recently been extended to protection of skin by topical cysteamine applied before exposure to helium ions (modal LET of 15 KeV per micron), with a resultant DMF of "at least 1.2" [23]. Preliminary observations by D. Baker (Mt. Zion Hospital, San Francisco, California) indicate that 5% cysteamine solution applied to the oral mucosa of rats 15 min before irradiation of brain tumors with 2000 rads reduced deaths due to mucosal damage from more than 50% in the unprotected animals to 0-10% [Baker, personal communication]. Since the radiation response of tumors growing beneath topically-protected tissues does not appear to be affected, this approach would seem to hold promise for clinical use.

Recommendation on protectors:

Further experimental work directed towards the following specific areas in radiation protection is indicated:

1. Further testing of WR-2721 (and related compounds), with particular emphasis on its efficacy against an expanded range of animal tumors at various sites and at various stages of growth. Multifraction, rather than single dose, experiments and late, as well as acute, tissue effects should be emphasized.

2. Continued screening for new agents. The recent development of at least two rapid and inexpensive in-vivo screening procedures [31,32] provides an alternative to expensive LD_{50} studies, at least for purposes of initial screenings.

3. Critical evaluation of topical thiol protection within the context of differential protection for tissues surrounding or overlying tumors.

4. Intensification of basic research into the mechanisms of chemical protection. This venerable subject has received new impetus from the development of new agents [33] and the use of "counter-protective" reagents such as diamide [27].

III. RECENT DEVELOPMENTS IN SELECTED AREAS OF CHEMICAL RADIOSENSITIZATION

That certain chemicals can sensitize biological systems to the effects of ionizing radiation has been known for many years, but it is only recently that this field has made much progress. In broad terms, sensitizing chemicals may act <u>before</u> irradiation (e.g., by synchronizing cells in a radiosensitive phase of the cell cycle), <u>during</u> irradiation (e.g., by interacting with radiation-induced radicals to amplify their effects), or <u>after</u> irradiation (e.g., by preventing repair of sublethal or potentially lethal damage). Each of these approaches has potential application to cancer therapy, but the context of this meeting would seem to require that we use the term "sensitizer" only in its narrowest sense -- that is, a chemical that decreases the survival of irradiated cells when present during, but not after, the radiation exposure. My usage of the term will conform to this convention.

Most of the recent work with sensitizers has focused on compounds that sensitize anoxic, but not euoxic, cells. Many of these compounds will be described and characterized in papers to be delivered at this meeting -- including the remainder of my own -- and it is clear to all of us that they have much to offer in terms of our basic knowledge about anoxic radioresistance. Before reviewing my own work in this field (I shall leave the so-called "electron-affinic" sensitizers for others to discuss), however, I wish to enter a strong plea that we not let our successes with anoxic sensitizers distract us from the very real need for agents which might act, also, on euoxic tumor cells (but not, of course, on normal tissue). The basic premise underlying the current enthusiasm for anoxic sensitizers is, after all, that anoxic cells exist within human tumors and that they constitute a significant barrier to successful radiotherapy. There is little or no direct evidence that this is true. Indeed, there is fairly good reason to believe that even if such cells do exist, the reoxygenation which probably occurs during fractionated therapy deals with them adequately. Considering this, one of the most eminent clinician/scientists in the United States wrote, as recently as 1974, that even if anoxic sensitizers should become available for clinical use, "...it is unlikely that the response of most tumors will be significantly altered" [34].

Having said this, however, I must hasten to admit that our only major breakthroughs in recent years have involved anoxic sensitizers and proceed to review for you our own contributions to this field. In marked contrast to the carefully thought-out studies carried out over the years by Adams and his colleagues, our studies of sensitization began almost by accident. We had set out to verify the then widely-accepted idea that cellular radioresistance, particularly under anoxic conditions, was caused by intracellular glutathione (GSH) content. Since GSH is known to be a reasonably good radioprotective agent and all mammalian cells contain the compound at concentrations as high as 5 mM, this idea seemed eminently reasonable, so we were quite surprised to find that cellular GSH levels did <u>not</u> correlate with radiosensitivity during the cell cycle [35] or in plateau-phase G_1 [36]. In an attempt to approach this problem more directly, we turned to a new GSH-oxidizing agent, diazene dicarboxylic acid bis (N,N'-dimethylamide) $[(CH_3)_2NCON=NCON(CH_3)_2]$, or "diamide" [37]. Initial studies indicated that although diamide reacts with a variety of redox compounds when present in excess [38,39], it specifically and reversibly oxidizes intracellular GSH when used at low concentrations (10-20 μM with $1-2\times10^6$ cells/ml) under carefully controlled conditions (Figure 2) [40].

When we tested the radiosensitivity of cells that had been treated with just enough diamide to oxidize their GSH, we were surprised to find that neither their euoxic nor their anoxic sensitivity was altered [41,42]. These results, some of which are shown in Figures 3 and 4, indicated that GSH does not play a major role in cellular radiation resistance, a conclusion that finds additional support in recent studies by other workers [43,44,44a]. In contrast, when we irradiated cells with higher concentrations of diamide anoxic cells were markedly sensitized, the most prominent feature of their response being a dramatic reduction of the shoulder [41-43]. This phenomenon is never observed with the "electron-affinic" sensitizers. Experiments with bacterial mutants [44b] and with rapid-mix techniques [43a] subsequently indicated that diamide sensitized anoxic cells by at least one mechanism not shared by oxygen or the "electron-affinic" sensitizers that mimic it. These studies did not otherwise succeed in identifying the mechanisms by which diamide sensitizes.

When we tested the efficacy of diamide as an in-vivo sensitizer, we were disappointed to find that it was completely ineffective (even with intra-tumor injection) against the solid EMT-6 mouse tumor, although it did sensitize ascites tumor cells [45]. In any event, the compound was far too reactive and toxic for in-vivo use.

Despite this disappointment, we continued our studies because diamide was the only anoxic sensitizer to affect the shoulder and we felt that it might provide significant insight into the mechanism of shoulder modification. (It should be pointed out that most of the fractionated radiation doses delivered in therapy fall on or near the shoulder region of the survival curve,

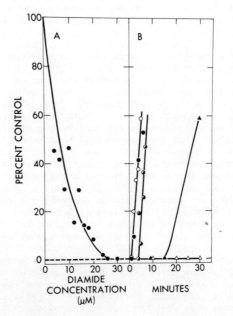

FIG.2. Oxidation and regeneration of endogenous non-protein sulfhydryl (NPSH) in Chinese hamster cells suspended in PBS and treated with diamide.

Panel A: Oxidation of NPSH in V79 cells at $0°C$. The reaction was stopped by adding cold sulfosalicylic acid 3–5 min after diamide was added.

Panel B: Regeneration of NPSH. V79 or CHO cells suspended in PBS at 2×10^6/ml were treated with 40μM diamide in the absence of glucose. Thirty seconds later, a concentrated glucose solution was added to some (final concentration, 5 mg/ml) and an equal volume of PBS was added to others. The cells were then incubated at the appropriate temperature for the times shown, after which ice-cold sulfosalicylic acid was added to stop regeneration. Symbols: (●) V79 and (○) CHO cells treated with 40μM diamide and incubated at 25°C with glucose; (◐) CHO cells treated with 80μM diamide and incubated at 25°C with glucose; (▲) V79 cells treated with 40μM diamide and incubated at 25°C without glucose; (△) V79 cells treated with 40μM diamide and incubated at 0°C without glucose. (From Ref. [46].)

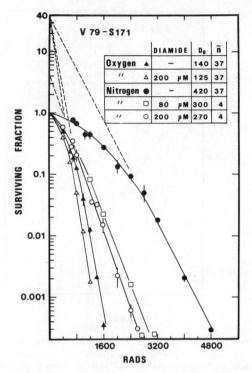

FIG.3. Effect of diamide on the survival of V79-S171 cells irradiated while suspended at 2×10^6/ml in ice-cold PBS. Each symbol (± SE when this exceeds the size of the symbol) represents 2 – 12 experiments except in the case of oxygen plus diamide (1 experiment). The OER for untreated cells is 3.00; survival parameters are shown on the figure. (From Ref. [42].)

so that sensitization in this dose range would have particular clinical interest.) At the same time, we initiated a search for analogues of diamide that might teach us something about the sensitizer configuration necessary to effect anoxic shoulder modification. Although these studies are of recent vintage, we are pleased to be able to report a modest degree of success on both fronts.

In our recent studies, (most of which are reported in detail elsewhere [42,46]), we examined the time course of radiosensitization by diamide, its effects on radiation-induced chromosome aberrations and DNA strand breaks, the correlation between cellular redox state and sensitization, and the ability of sensitized cells to repair sublethal damage. The major results of these studies may be summarized as follows:

1. low concentrations of diamide (40 µM for Chinese hamster ovary cells) remove the shoulder from the anoxic survival curve, whereas higher concentrations also decrease the D_0 (Figure 5). Euoxic cells are sensitized slightly, but only at high concentrations and only by decreasing D_0;

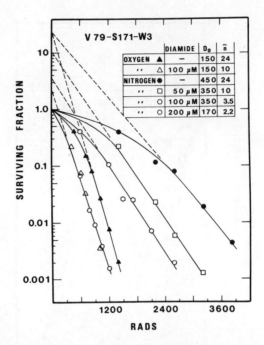

FIG.4. Effect of diamide on the survival of V79-S171-W3 cells irradiated as monolayers in complete medium at 18°C. Hypoxia in this experiment was obtained by repeated flushing of aluminium chambers containing the culture dishes with nitrogen. Each dish contained $10^2 - 10^4$ cells, depending on the anticipated survival level. Diamide was added 2.5 h before irradiation. (From Ref.[42].)

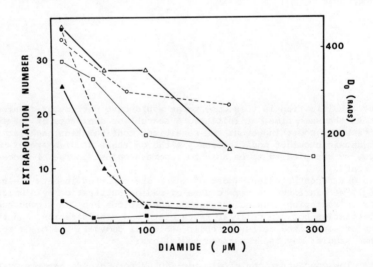

FIG.5. Effects of diamide on extrapolation number (▲, ●, ■) and D_0 (△, ○, □) of V79 and CHO cells irradiated in nitrogen. (△, ▲) V79-S171-W3 cells irradiated as monolayers in complete medium at 18°C; (○, ●) V79-S171 cells irradiated in ice-cold PBS; (□, ■) CHO spinner cells irradiated in ice-cold PBS.

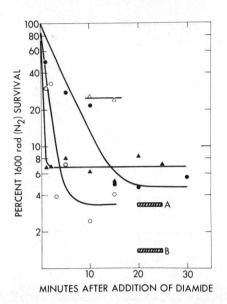

FIG.6. *Time-course for the development of diamide radiosensitization in hypoxic V79 cells. Cells (2×10^6/ml PBS) were gassed with nitrogen for 20 min and diamide was then added by a microdroplet technique described in [46]. At the times indicated, a test dose of 1600 rad was delivered and the cells were then plated for colony formation. Symbols: (●) 200µM diamide, 0°C, no glucose; (○) 200µM diamide, 25°C, no glucose; (△) 200µM diamide, 25°C, 2 mg glucose/ml; (A) 200µM diamide, 0°C and 25°C, no glucose; (▲) 500µM diamide, 25°C, 2 mg glucose/ml; (B) 500µM diamide, 0°C and 25°C, no glucose. (From Ref. [46].)*

2. sensitization of anoxic mammalian cells by 200 µM diamide requires 10-15 min to develop in the cold, although the sensitizer enters the cells and oxidizes GSH within seconds (Figure 6) [46]. Delayed sensitization was not observed when DNA strand breaks were measured instead of cell survival, indicating that the sensitizer penetrated to the DNA within a short period of time, nor was it observed at 20° C (Figure 6).
3. diamide induced significantly less sensitization in the presence of glucose (20° C) than in its absence (Figure 6). This could not be accounted for by metabolic inactivation of sensitizer molecules.
4. cells treated with diamide and then washed free of the sensitizer and irradiated at 0° C are sensitized to the same extent as if diamide was still present, even though they are demonstrably free of any intracellular or extracelluar sensitizer molecules [unpublished data]. This effect is reversed within seconds if the cells are permitted to metabolize exogenous or endogenous substrates.
5. diamide does not interfere with split-dose repair (Figure 7), although it does prevent repair if left on the cells between the two radiation doses.
6. diamide sensitization does not correlate with GSH levels but does show some correlation with cellular NADH + NADPH content [46].

We have interpreted these results as indicating that diamide sensitizes anoxic cells by at least two mechanisms, one involving interference with a rapid repair mechanism and the other involving direct interaction with radiation-induced DNA radicals [46]. The data strongly suggest that diamide-induced oxidation of pyridine nucleotides is involved in the first of these (associated with shoulder modification), although the details remain obscure.

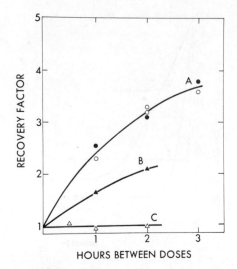

FIG. 7. *Effects of diamide on repair of sublethal damage by irradiated V79 cells.*

Panel A: Hypoxic irradiation in diamide followed by repair in air: cells (2×10^6/ml) were suspended in PBS containing 200µM diamide ($0°C$), made hypoxic, and irradiated with 800 rad. They were then plated in complete medium without diamide at $37°C$ for 1–3 h and given a second radiation dose (700 rad in air). Diamide-treated but unirradiated controls were included, as were cells that received 800 rads euoxically in the absence of sensitizer. (●) First dose hypoxic in diamide, second dose euoxic; (○) both doses euoxic.

Panels B and C: Irradiation in air without diamide followed by repair in the presence (△) or absence (▲) of diamide: cells (subline W3) growing on 60-mm petri dishes were irradiated euoxically at $18°C$, incubated for various times at $18°C$, and then given a second radiation dose. Diamide, when present, remained in the medium during the entire period, including both irradiations; the initial concentration was 100µM. Radiation doses were 275 rad each for cells in diamide and 550 rad each for cells without diamide; these gave similar survival levels when no interval separated the doses. (From Ref. [46].)

It is of interest, in this context, that metronidazole (Flagyl), which does not affect the survival curve shoulder [47], also does not oxidize pyridine nucleotides [48]. This compound is also unreactive with sulfhydryls in intact cells (unpublished data) although such reactions have been observed in an iron-catalyzed in-vitro model [49].

In other studies, conducted in collaboration with Drs. E. and N. Kosower in Tel Aviv, we have recently discovered that a diamide-like compound, diazene dicarboxylic acid bis (N'-methylpiperazide), or "DIP" [50]

$$CH_3N\underset{}{\bigcirc}NCON = NCON\underset{}{\bigcirc}NCH_3$$

mimics diamide in removing the shoulder from the anoxic survival curve [51]. This compound does not affect the anoxic D_0 (Figure 8). Another, equally reactive, analogue, DIP+2 [50], does not cross the cell membrane and does not sensitize anoxic cells [51]. Since the rate at which DIP penetrates cells varies with temperature, [50,51], this compound provides a useful probe for exploration of the mechanism(s) of anoxic shoulder modification.

We are also continuing our studies of sensitizer combinations, and in particular combinations of diamide with various of the electron-affinic sensitizers. Early results [52] indicate that non-sensitizing concentrations of

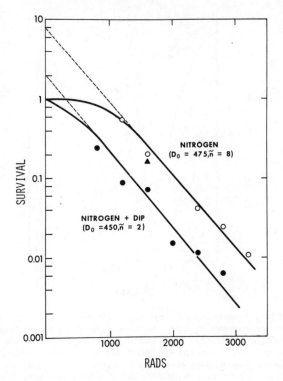

FIG.8. Survival of V79 Chinese hamster cells irradiated hypoxically in the presence or absence of DIP or DIP+2. Cells (2×10^6/ml in PBS at $18°C$) were made hypoxic by gassing them with nitrogen for 20 min. DIP or DIP+2 (200 nmoles/2×10^6 cells) was then added without breaking the seal and X-radiation was begun 4 min later. The slopes of the survival curves (D_0) and the extrapolation numbers ($\tilde{n}$) are indicated on the figure. (○) untreated (●) DIP-treated, (▲) DIP+2-treated. (From Ref.[51].)

diamide or nifuroxime, when present together during irradiation of Chinese hamster cells, sensitize synergistically (Figure 9). Indeed, by using higher concentrations, we have been able to sensitize anoxic cells better than with oxygen itself. We believe that this effect represents removal, by diamide, of cellular molecules that normally would either antagonize the action of nifuroxime or inactivate it. This approach deserves further study, particularly as a means of obtaining sensitization in patients at drug concentrations low enough to avoid acute toxic reactions.

Finally, I wish to reiterate my earlier plea that we not become affected by "tunnel vision." Even in the case of the electron-affinic sensitizers that have progressed to the point of clinical trials [53,54], we must keep in mind that we still have much to learn -- for example, about their "non-radiation" effects such as increased oxygen diffusion [55] and selective cytotoxicity [56, 57]. The possibility that such effects may occur in vivo should be kept in mind when clinical radiotherapy trials are designed and evaluated. We must also continue the search for sensitizers that are effective against euoxic tumor cells. To date, only the halogenated pyrimidines [58] have found a significant niche in clinical radiotherapy, a certain degree of success having been obtained with these compounds in certain brain tumors [59] and, more recently, in a few cases of osteosarcoma [60].

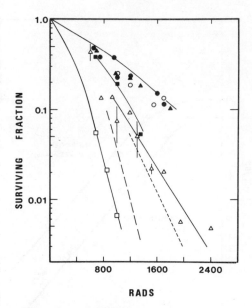

FIG.9. *Survival of V79-S171 cells irradiated in the presence of diamide, nifuroxime, or both together. (●) Nitrogen only, (▲) 20μM nifuroxime in nitrogen, (○) 20μM diamide in nitrogen, (△) 20μM each of diamide and nifuroxime in nitrogen, (■) 125μM nifuroxime, (□) 200μM diamide plus 125μM nifuroxime in nitrogen. Survival curves for cells irradiated in oxygen (— —) or in nitrogen with 200μM diamide (---) are included for comparison. Since the nitrogen control points from the present experiments fell slightly below our usual hypoxic curve, that curve is reproduced on the figure in place of a curve fitted to the actual experimental points. (From Ref. [52].)*

Recommendations on sensitizers:

Intensive experimental work is required in the following areas:

1. Expansion of the search for more effective and less toxic sensitizers, both those that sensitize anoxic cells and those that sensitize on other bases. In view of the fact that 90% of the present testing is limited to two or three lines of cultured Chinese hamster cells, it seems important that other types of cells (particularly human) be examined. In-vivo testing must include as wide a variety of tumor types and growth rates as possible [cf. 61] and should emphasize multifraction, rather than single dose, exposures.

2. Combination studies, particularly with sensitizers that act by independent mechanisms, should be expanded. The possibility that chemical protection of normal tissues (e.g., with WR-2721) might be combined with anoxic sensitizers is particularly enticing and should be evaluated; preliminary studies along these lines have yielded encouraging results [J. Yuha personal communication].

3. Examination of the effects of sensitizers on host immunity, particularly insofar as these might bear on failure of local control or growth of metastases. There is recent evidence, for example, that niridazole, which is a weak radiosensitizer [47], suppresses cellular immunity in humans [62]. Other potential hazards of sensitizers, such as mutagenicity [63-65], carcinogenicity [66], and genetic or metabolic predisposition to drug-induced hemolysis (known, for example, to be a problem with some nitrofurans), should also be evaluated and quantitated.

4. Evaluation of the effects, if any, of sensitizer treatment on the response of normal and malignant tissues to chemotherapeutic agents which might be employed, together with radiation, in treating particular malignancies.

IV. CONCLUSION

Although radiation modifiers have yet to find a significant place in the therapeutic armamentarium, recent developments provide good reason for cautious optimism about the future. Cancer affects millions of human beings -- in the United States alone, 610 000 new cases are diagnosed each year [67] -- and a significant proportion of those cases in which radiotherapy fails to establish local tumor control could be treated successfully if only slightly higher doses could be delivered or if the doses now used could be made more effective. The potential contribution of chemical modifiers to this situation is obvious. Even if we do not succeed in increasing cure rates for most cancers, it may be possible to make a contribution in terms of decreased acute and late tissue reactions that could ease the suffering and disability of the cancer patient. From even this brief review of the field of chemical modification it is clear that there is much to do -- but it is even clearer, to all of us, that the potential benefits are well worth any efforts we can make.

ACKNOWLEDGEMENTS

The author's research is supported by the United States Energy Research and Development Administration and is made possible by the outstanding collaboration of many colleagues, in particular J. Power and C. Koch. This paper is dedicated to Mary Harris, whose personal struggle with cancer provides constant inspiration.

REFERENCES

[1] FOYE, W.O., Radiation-protective agents in mammals, J. Pharm. Sci. 58 (1969) 283.
[2] KLAYMAN, D.L., COPELAND, E.S., "Design of antiradiation agents", Ch 2, Drug Design (VI), (ARIENS, E.J., Ed), Academic Press, New York (1975).
[3] YUHAS, J.M., STORER, J.B., Differential chemoprotection of normal and malignant tissues, J. Natl. Cancer Inst. 42 (1969) 331.
[4] RAPPLEYE, A.T., JOHNSON, G.H., OLSEN, J.D., LAGASSE, L.D., The radioprotective effects of vasopressin on the gastrointestinal tract of mice, Radiol. 117 (1975) 199.
[5] PREWITT, R.L., MUSACCHIA, X.J., Mechanisms of radio-protection by catecholamines in the hamster (Mesocricetus auratus), Int. J. Radiat. Biol. 27 (1975) 181.
[6] STECKEL, R.J., et al., Radiation protection of the normal kidney by selective arterial infusions, Cancer 34 (1974) 1046.
[7] YUHAS, J.M., STORER, J.B., Chemoprotection against three modes of radiation death in the mouse, Int. J. Radiat. Biol. 15 (1969) 233.
[8] YUHAS, J.M., Radioprotective and toxic effects of S-2-(3-aminopropylamino) ethylphosphoro thioic acid (WR-2721) on the development of immunocompetent cells, Cell. Immunol. 4 (1972) 256.
[9] YUHAS, J.M., Radiotherapy of experimental lung tumors in the presence and absence of a radioprotective drug, S-2-(3-aminopropylamino) ethylphosphorothioic acid (WR-2721), J. Natl. Cancer Inst. 50 (1973) 69.

[10] HARRIS, J.W., PHILLIPS, T.L. Radiobiological and biochemical studies of thiophosphate radioprotective compounds related to cysteamine, Radiat. Res. 46 (1971) 362.

[11] PHILLIPS, T.L., KANE, L., UTLEY, J.F., Radioprotection of tumor and normal tissues by thiophosphate compounds, Cancer 32 (1973) 528.

[12] SIGDESTAD, C.P., CONNOR, A.M., SCOTT, R.M., The effect of S-2-(3-aminopropylamino)ethylphosphorothioic acid (WR-2721) on intestinal crypt survival. I. 4 MeV X-rays, Radiat. Res. 62 (1975) 267.

[13] LOWY, R.O., BAKER, D.G., Protection against local irradiation injury to the skin by locally and systemically applied drugs, Radiol. 105 (1972) 425.

[14] LOWY, R.O., BAKER, D.G., Effect of radioprotective drugs on the therapeutic ratio for a mouse tumor system, Acta Radiol. 12 (1973) 425.

[15] UTLEY, J.F., PHILLIPS, T.L., KANE, L.J., WHARAM, M.D., WARA, W.M., Differential radioprotection of euoxic and hypoxic mouse mammary tumors by a thiophosphate compound, Radiol. 110 (1974) 213.

[16] YUHAS, J.M., Biological factors affecting the radioprotective efficiency of S-2-[3-aminopropylamino]ethylphosphorothioic acid (WR-2721). $LD_{50(30)}$ doses, Radiat. Res. 44 (1970) 621.

[17] YUHAS, J.M., PROCTOR, J.O., SMITH, L.H., Some pharmacologic effects of WR-2721: their role in toxicity and radioprotection, Radiat. Res. 54 (1973) 222.

[18] CALDWELL, R.W., HEIFFER, M.H., Acute cardiovascular and autonomic effects of WR-2721: a radioprotective compound, Radiat. Res. 62 (1975) 62.

[19] WASHBURN, L.C., CARLTON, J.E., HAYES, R.L., YUHAS, J.M., Distribution of WR-2721 in normal and malignant tissues of mice and rats bearing solid tumors: dependence on tumor type, drug dose and species, Radiat. Res. 59 (1974) 475.

[20] KOLLMAN, G., MARTIN, D., SHAPIRO, B., The distribution and metabolism of the radiation protective agent aminopentylaminoethylphosphorothioate in mice, Radiat. Res. 48 (1971) 542.

[21] UTLEY, J.F., PHILLIPS, T.L., KANE, L.J., Protection of normal tissues by WR-2721 during fractionated irradiation, Int. J. Radiat. Oncol. Biol. Phys. (in press).

[22] SIGDESTAD, C.P., CONNOR, A.M., SCOTT, R.M., The effect of S-2-(3-aminopropylamino)ethylphosphorothioic acid (WR-2721) on intestinal crypt survival. II. Fission neutrons, Radiat. Res. (in press).

[23] BAKER, D.G., LEITH, J.T., Protection of the skin of mice against irradiation with cyclotron-accelerated helium ions by 2-mercapto-ethylamine, Acta Radiol. (in press).

[24] MARTIN, D., KOLLMAN, G., SHAPIRO, B., The radiation decomposition of phosphorothioate protective agents, Radiat. Res. 56 (1973) 246.

[25] HOROWITZ, S.B., FENICHEL, I.R., HOFFMAN, B., KOLLMAN, G., SHAPIRO, B., The intracellular transport and distribution of cysteamine phosphate derivatives, Biophys. J. 10 (1970) 994.

[26] HAHN, A., LOHMANN, W., HILLERBRAND, M., DEFFNER, U., Molecular mechanism of action of the radioprotective substance WR-2721, Radiat. Envir. Biophys. 11 (1975) 265.

[27] POWER, J.A., GOLDSTEIN, L.S., HARRIS, J.W., A test of the "mixed disulphide" hypothesis of cysteamine radioprotection, Int. J. Radiat. Biol. 26 (1974) 91.

[28] ANTONE, H.J., GIBBS, S.J., Chemical protection of mammalian oral mucosa exposed to daily local radiation, J. Dent. Res. 52 (1973) 1153.

[29] GOEPP, R., FITCH, F., Parenteral chemical protection against oral radiation death in mice, Radiat. Res. 31 (1967) 149.

[30] GOEPP, R., FITCH, F., Topical chemical protection against oral radiation death in mice, Radiat. Res. 34 (1968) 36.

[31]   HEDDLE, J.A., HARRIS, J.W., Rapid screening of radioprotective drugs in vivo, Radiat. Res. 61 (1975) 350.
[32]   WREDE, D.E., CHEN, I.W., KEREIAKES, SAENGER, E.L., Evaluation of radiochemical protectors using urinary deoxycytidine levels, Int. J. Radiat. Biol. 28 (1975) 117.
[33]   HART, R.W., GIBSON, R.E., CHAPMAN, J.D., REUVERS, A.P., SINHA, B.K., GRIFFITH, R.K., WITIAK, D.T., A radioprotective stereostructure-activity study of cis- and trans-2-mercaptocyclobutylamine analogs and homologs of 2-mercaptoethylamine, J. Med. Chem. 18 (1975) 323.
[34]   KAPLAN, H.S., On the relative importance of hypoxic cells for the radiotherapy of human tumours, Eur. J. Cancer 10 (1974) 275.
[35]   HARRIS, J.W., TENG, S.S., Sulfhydryl groups during the S phase: Comparison of cells from G_1, plateau-phase G_1 and G_0, J. Cell. Physiol. 81 (1973) 91.
[36]   HARRIS, J.W., PAINTER, R.B., HAHN, G.M., Endogenous non-protein sulfhydryl and cellular radiosensitivity, Int. J. Radiat. Biol. 15 (1969) 289.
[37]   KOSOWER, N.S., KOSOWER, E.M., WERTHEIM, B., CORREA, W.S., Diamide, a new reagent for the intracellular oxidation of glutathione to the disulfide, Biochem. Biophys. Res. Commun. 37 (1969) 593.
[38]   HARRIS, J.W., BIAGLOW, J.E., Non-specific reactions of the glutathione oxidant "diamide" with mammalian cells, Biochem. Biophys. Res. Commun. 46 (1972) 1743.
[39]   KOSOWER, E.M., CORREA, W., KINON, B.J., KOSOWER, N.S., Glutathione. VIII. Differentiation among substrates by the thiol-oxidizing agent, diamide, Biochim. Biophys. Acta 264 (1972) 39.
[40]   HARRIS, J.W., ALLEN, N.P., TENG, S.S., Evaluation of a new glutathione-oxidizing reagent for studies of nucleated mammalian cells, Exptl. Cell Res. 68 (1971) 1.
[41]   HARRIS, J.W., POWER, J.A., Diamide: a new sensitizer for anoxic cells, Radiat. Res. 56 (1973) 97.
[42]   HARRIS, J.W., POWER, J.A., KOCH, C.J., Radiosensitization of hypoxic mammalian cells by diamide. I. Effects of experimental conditions on survival, Radiat. Res. (in press).
[43]   VAN HEMMEN, J.J., MEULING, W.J.A., BLEICHRODT, J.F., Radiosensitization of biologically active DNA in cellular extracts by oxygen. Evidence that the presence of SH-compounds is not required, Int. J. Radiat. Biol. 26 (1974) 547.
[43a] WATTS, M.E., WHILLANS, D.W., ADAMS, G.E., Studies of the mechanisms of radiosensitization of bacterial and mammalian cells by diamide, Int. J. Radiat. Biol. 27 (1975) 259.
[44]   HO, Y.L., HO, S.K., Studies of parachloromercuribenzoate-induced radiosensitization in Escherichia coli and bacteriophages, Radiat. Res. 61 (1975) 230.
[44a] APONTOWEIL, P., BERENDS, W., Isolation and characterization of glutathione-deficient mutants of Escherichia coli K12, Biochim. Biophys. Acta 399 (1975) 10.
[44b] GOLDSTEIN, L.S., HARRIS, J.W., Radiosensitization of anoxic Escherichia coli mutants by diamide, Int. J. Radiat. Biol. 25 (1974) 391.
[45]   HARRIS, J.W., WARA, W.M., KANE, L.J., Sensitization of anoxic mouse tumours to X-rays with diamide and nifuroxime, Int. J. Radiat. Biol. 26 (1974) 227.
[46]   HARRIS, J.W., BIAGLOW, J.E., KOCH, C.J., POWER, J.A., Radiosensitization of hypoxic mammalian cells by diamide. II. Studies of mechanism, (submitted for publication).
[47]   CHAPMAN, J.D., REUVERS, A.P., BORSA, J. Effectiveness of nitrofuran derivatives in sensitizing hypoxic mammalian cells to X-rays, Brit. J. Radiol. 46 (1973) 623.

[48] COOMBS, G.H., RABIN, B.R., Flagyl and reduced NAD, FEBS Lett. 42 (1974) 105.
[49] WILLSON, R.L., SEARLE, S.J.F., Metronidazole (Flagyl): iron catalysed reaction with sulphydryl groups and tumor radiosensitization, Nature 255 (1975) 498.
[50] KOSOWER, E.M., KOSOWER, N.S., KENETY-LONDNER, H., LEVY, L., Glutathione IX. New thiol-oxidizing agents: DIP, DIP+1, DIP+2, Biochem. Biophys. Res. Commun. 59 (1974) 347.
[51] HARRIS, J.W., POWER, J.A., KOSOWER, N.S., KOSOWER, E.M., DIP and DIP+2 as glutathione oxidants and radiation sensitizers in cultured Chinese hamster cells, Int. J. Radiat. Biol. (in press).
[52] HARRIS, J.W., POWER, J.A., KOCH, C.J., Synergistic chemical radiosensitization of hypoxic mammalian cells, Int. J. Radiat. Oncol. Biol. Phys. (in press).
[53] FOSTER, J.L., et al., Serum concentration measurements in man of the radiosensitizer Ro-07-0582: some preliminary results, Br. J. Cancer 31 (1975) 679.
[54] URTASUN, R.C., et al., Phase I study of high-dose metronidazole: a specific in vivo and in vitro radiosensitizer of hypoxic cells, Radiol. 117 (1975) 129.
[55] BIAGLOW, J.E., DURAND, R.E., The effects of nitrobenzene derivatives on oxygen utilization and radiation response of an in vitro tumor model, Radiat. Res. (in press).
[56] SUTHERLAND, R.M., Selective chemotherapy of noncycling cells in an in vitro tumor model, Cancer Res. 34 (1974) 3501.
[57] OLIVE, P.L., MCCALLA, D.R., Damage to mammalian cell DNA by nitrofurans, Cancer Res. 35 (1975) 781.
[58] SZYBALSKI, W., X ray sensitization by halopyrimidines, Canc. Chemother. Repts. 58 (1974) 539.
[59] HOSHINO, T., SANO, K., Radiosensitization of malignant brain tumors with bromouridine (thymidine analogue), Acta Radiol. 8 (1969) 15.
[60] GOFFINET, D.R., KAPLAN, H.S., DONALDSON, S.S., BAGSHAW, M.A., WILBUR, J.R., Combined radiosensitizer infusion and irradiation of osteogenic sarcoma, Radiol. 117 (1975) 211.
[61] HOWLETT, J.F., THOMLINSON, R.H., ALPER, T. A marked dependency of the comparative effectiveness of neutrons on tumour line and its implications for clinical trials, Brit. J. Radiol. 48 (1975) 40.
[62] WEBSTER, L.T., BUTTERWORTH, A.E., MAHMOUD, A.A.F., MNGOLA, E.N., WARREN, K.S., Suppression of delayed hypersensitivity in schistosome-infected patients by niridazole, New Engl. J. Med. 292 (1975) 1144.
[63] MCCALLA, D.R., VOUTSINOS, D., On the mutagenicity of nitrofurans, Mutation Res., 26 (1974) 3.
[64] VOOGD, C.E., VAN DER STEL, J.J., JACOBS, J. The mutagenic action of nitroimidazoles. I. Metronidazole, nimorazole, dimetridazole and ronidazole, Mutation Res. 26 (1974) 483.
[65] VOOGD, C.E., VAN DER STEL, J.J., JACOBS, J., The mutagenic action of nitroimidazoles. II. Effects of 2-nitroimidazoles, Mutation Res. 26 (1975) 149.
[66] COHEN, S.M., ERTÜRK, E., VON ESCH, A.M., CROVETTI, A.J., BRYAN, G.T., Carcinogenicity of 5-nitrofurans and related compounds with amino-hetero-cyclic substitutes, J. Natl. Cancer Inst. 54 (1975) 841.
[67] CULTER, S.J., SCOTTO, J., DEVESA, S.S., CONNELLY, R.R., Third national cancer survey -- an overview of available information, J. Natl. Canc. Inst. 53 (1974) 1565.

RADIATION CHEMICAL BASIS FOR THE ROLE OF GLUTATHIONE IN CELLULAR RADIATION SENSITIVITY*

M. QUINTILIANI, R. BADIELLO, M. TAMBA
Laboratorio di Fotochimica e Radiazioni d'Alta Energia,
Bologna, Italy

and

G. GORIN
Chemistry Department,
Oklahoma State University,
Stillwater, Oklahoma,
United States of America

Abstract

RADIATION CHEMICAL BASIS FOR THE ROLE OF GLUTATHIONE IN CELLULAR RADIATION SENSITIVITY.

It is widely believed that glutathione, the most abundant endogenous low-molecular-weight sulphydryl compound, plays a part in modulating radiation sensitivity of living cells. Moreover, the radioprotection afforded by exogenous sulphydryl compounds has been interpreted, by some authors, in terms of intracellular release of reduced glutathione. The literature and the authors' data on steady-state and pulse radiolysis of glutathione in reduced and in oxidized form show that in aqueous solution practically all the relevant events occur on the cysteine moiety of the molecule, the SH group being the focal point of chemical changes. The very rapid rates of reaction of the free radicals from water with reduced and oxidized glutathione, the ability of the SH compound to operate free radical repair by hydrogen transfer and that of forming mixed disulphides with proteins still remain the most plausible mechanisms to explain the role of glutathione in affecting biological radiation sensitivity.

INTRODUCTION

For more than twenty years a great deal of interest and discussion has been devoted to the role of SH and SS compounds in determining the radiation sensitivity of biological systems. The discovery that SH compounds are possibly the most active and the most versatile radioprotectors, together with the understanding of the important biological functions displayed by sulphydryls and disulphides in the dynamics of living matter, are at the origin of such interest and discussion.

Without going into an historical review, the up-to-date state of knowledge can be summarized as follows:

- SH compounds and, in some cases also SS compounds, protect biological systems from radiation damage in a way which, as yet, has not been fully understood [1];
- since exogenous sulphydryls protect living cells from radiation damage, it is reasonable to assume that endogenous sulphydryls, in particular the non-protein-bound sulphydryls (NPSH), which naturally occur in all cells and tissues, affect the radiation sensitivity of cells and organisms;

---

* This paper originates from a Co-operative Research Programme supported by the Consiglio Nazionale delle Ricerche, Italy and the National Science Foundation, USA.

- as glutathione is the most abundant intracellular low-molecular-weight thiol [2], it can be assumed that the eventual role of modulating cellular radiation sensitivity should be attributed to this compound;
- the protection afforded by exogenous sulphydryls should also be attributed to intracellular glutathione, according to the hypothesis proposed by Révész and Modig [3] who postulated that exogenous thiols release glutathione from binding sites on cellular proteins.

The present paper is an attempt to discuss the problems outlined in these statements in relation to radiation chemistry of glutathione.

BIOLOGICAL INDICATIONS ON THE ROLE OF GLUTATHIONE IN RADIATION SENSITIVITY

Positive experimental evidence supporting the importance of NPSH in cellular radiation sensitivity is far from overwhelming. Some pertinent observations have been reported about mammalian cells. Caspersson and Révész [4] showed in 1963 that a radioresistant line of ascites tumour cells had a higher content of sulphydryl compounds than a radiosensitive line. In 1969, Ohara and Terasima [5] found in HeLa cells a rather remarkable, almost 1:1, correspondence between variations in radiation survival and in cellular NPSH concentration during the cell cycle. It is known that the lethal effect of radiation can vary according to the stage of the cell cycle at the moment of irradiation. This phenomenon has been extensively studied in synchronized cultures of mammalian cells, particularly by Sinclair [6]. While there are some notable exceptions, it appears, in general, that there is a close association of an increase of the radioresistance with the onset of DNA synthesis during the S period. The mitotic period coincides with the state of maximum radiosensitivity and, in cell lines with a long G_1, a second resistant phase is displayed in early G_1. Sinclair [7] has attempted to relate the radiation sensitivity to the rate of DNA synthesis and has found difficulties in fitting different cell lines into a satisfactory model. However, according to this author, some of these difficulties disappear if one considers a two-part model in which the rate of DNA synthesis, or some concomitant factor dependent upon it, is one of the factors controlling radiation sensitivity and at least one other factor, also varying during the S period, interplays with the previous one in determining the actual response to radiation. Sinclair has indicated this additional factor as factor Q, and has proposed that its nature is that of some form of sulphydryl, probably NPSH, that is to say, mostly unbound glutathione.

In studying the influence of low concentrations of N-ethylmaleimide (NEM) on radiation survival of oxygenated Chinese hamster cells, Sinclair [7] came first to the conclusion that the sensitizing effect, observed in these conditions, was due to NEM binding to cellular NPSH compounds. More recently he has changed his view [8], inclining rather to attribute the sensitizing effect to the inhibition of repair of lethal damage. According to this view, the possibility exists that NEM acts by blocking some SH-containing repair enzymes which fluctuate in quantity and availability during the cell cycle.

The literature also contains a number of negative results on the role of NPSH, and of glutathione, in affecting cellular radiosensitivity. For instance, Harris et al. [9] have shown that Chinese hamster cells in plateau growth contain only one third the amount of NPSH of exponentially growing cells, yet the radiosensitivity of the two is identical. More recent data along this line have been produced by Harris and Power [10] using a compound: diamide (diazenedicarboxylic acid bis (N,N-dimethylamide)), which, under appropriate conditions, is able to very rapidly oxidize the GSH of intact mammalian cells to the corresponding disulphide. In concentrations not far exceeding those necessary to oxidize all the NPSH, diamide did not affect radiation sensitivity of anoxic or euoxic mammalian cells. When in large excess, diamide sensitized only anoxic cells. Similar results were obtained in bacteria, where diamide, even in large excess, did not succeed in oxidizing all the intracellular NPSH. These results tend to show, on the one hand, that endogenous NPSH might not have such a critical role in cellular radiosensitivity, and, on the other hand, that the anoxic radiosensitization by relatively high concentrations of diamide should involve either the oxidation of protein thiol groups or some entirely different mechanism, like for instance some free radical reaction in which unreacted diamide is taking part. It should be recalled at this point that disulphides, which are formed by the action of diamide, are not believed to exert any protective action on cells. Such an assumption is based on experimental evidence concerning exogenous disulphides [11] which, presumably, do not penetrate mammalian cell membrane [12]. No data exist, as far as the authors know, on intracellular disulphides.

Some authors have postulated that endogenous NPSH only affect the radiosensitivity of anoxic cells, like Modig et al. [13] who pointed out that the sensitivity differences displayed by their tumour cell lines, with different NPSH content, when irradiated in anoxia, disappeared on irradiation in the presence of oxygen.

The hypothesis that intracellular thiols affect the radiosensitivity of anoxic cells could be nicely correlated with the observations of Howard-Flanders et al. [14, 15] and of Hutchinson [16]. These authors have shown that the radiosensitivity of biological macromolecules, like phage and transforming DNA or some enzymes, was not affected by oxygen when irradiated in vitro in purified solutions, but if a sulphydryl compound was added, a clear sensitizing effect of oxygen appeared. The oxygen effect could also be observed if the macromolecules were irradiated within the cells. Howard-Flanders et al. have suggested that a competition exists between oxygen and sulphydryl for radical sites produced by radiation in the macromolecules, resulting in permanent inactivation when the oxygen reacts and in restoration by hydrogen donation when sulphydryl reacts.

Howard-Flanders et al. [14] have also postulated that the reaction rate of oxygen with transforming DNA radical is 30 times faster than that of sulphydryl.

While these data are certainly suggestive, they are in some contradiction with those obtained in experiments where the oxygen dependence of the radioprotective effects of exogenous sulphydryls were studied. Both in microorganisms [17] and in mammalian cells [18] it was found that the highest protective effect could be measured when cells were irradiated in the presence of oxygen.

POSSIBLE MECHANISMS FOR THE ACTION OF GLUTATHIONE ON RADIOSENSITIVITY

The possible mechanisms whereby thiol compounds protect against radiation damage or affect natural radiation sensitivity are likely to operate either at the level of fast radiation chemical processes involving free radical reactions, leading to damage of critical molecules, or at the level of metabolic or biochemical processes which follow the previous phase and bring about the biological expression of the chemical damage.

In this paper the attention is focused mainly on radiation chemical mechanisms not only because this is the purpose of the paper, but also because no decisive evidence has been so far produced to show that the effects at the level of metabolic and biochemical processes are the relevant events for the increase in radioresistance.

The radiation chemical mechanisms which appear to be plausible have been proposed since several years and are still under discussion. They are: the radical scavenging, the hydrogen transfer and the formation of mixed disulphides.

It is believed that there is no need to illustrate them here since this has been done in many publications including a number of extensive reviews [19,20]. They are discussed in relation to radiation chemical data on glutathione in the final part of the paper.

THE RADIATION CHEMISTRY OF GLUTATHIONE

Steady-state radiolysis of aqueous solutions of glutathione, either in reduced or in oxidized form, shows that sulphur is the principal site of attack. The products which are formed are analogous to those originated in the radiolysis of solutions of cysteine or cystine, with no evidence for scission of the peptide chain. Only Bonotto and Netrawali [21] have reported that irradiation of oxygenated solutions of GSH resulted in some minor degradation of the peptide chain.

The major products of radiolysis of air-free aqueous solutions of GSH are: hydrogen, hydrogen sulphide, GSSG and γ-glutamylalanyl-glycine [22]. In analogy to the radiolysis of cysteine [23], that of GSH in the absence of oxygen can be described by the following reactions:

(1) $H_2O \longrightarrow OH^{\cdot} + H^{\cdot} + e^{-}_{aq} + H^{+} + H_2O_2 + H_2$

(2) $OH^{\cdot} + GSH \longrightarrow GS^{\cdot} + H_2O$

(3) $H^{\cdot} + GSH \longrightarrow GS^{\cdot} + H_2$

(4) $H^{\cdot} + GSH \longrightarrow G^{\cdot} + H_2S$

(5) $e^{-}_{aq} + H^{+} \longrightarrow H^{\cdot}$

(6) $e^{-}_{aq} + GSH \longrightarrow G^{\cdot} + HS^{-}, (\xrightarrow{+H^{+}} H_2S)$

(7) $G^{\cdot} + GSH \longrightarrow GH + GS^{\cdot}$

(8) $GS^{\cdot} + GS^{-} \rightleftharpoons GSSG^{-}$

(9) $GS^{\cdot} + GS^{\cdot} \longrightarrow GSSG$

(10) $GS^{\cdot} + GSSG^{-} \longrightarrow GSSG + GS^{-}$

(11) $H_2O_2 + 2 GSH \longrightarrow GSSG + 2 H_2O$

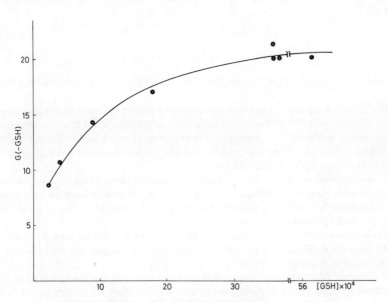

FIG.1. G(−GSH) for radiolysis of glutathione in the presence of oxygen.

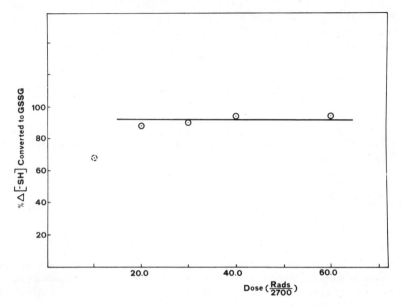

FIG.2. *Radiolytic conversion of GSH to GSSG. A plot of % △ [−SH] converted to GSSG against dose.* $[GSH] = 3 \times 10^{-3}$.

The G(-GSH) has been determined to be about 8 both for cysteine and glutathione [22, 23], a value which very well agrees with the theoretical value calculated for the reaction sequence given here.

Pulse radiolysis has contributed to the elucidation of the process by allowing one to measure the rate constants of many reactions and to identify some of the transients involved [22] like, for instance, the well-known radical anion RSSR⁻.

In our experiments we have measured the following rate constants:

$k_{2\,(OH\,+\,GSH)}$ = 1.3 × 10¹⁰ · M⁻¹ · sec⁻¹ at pH5
          1.7 × 10¹⁰ · M⁻¹ · sec⁻¹ at pH9

$k_{6\,(e^-_{aq}\,+\,GSH)}$ = 6.1 × 10⁹ · M⁻¹ · sec⁻¹ at pH7

$k_{8\,(GS^\bullet\,+\,GS^-)}$ = 6.6 × 10⁸ · M⁻¹ · sec⁻¹ at pH8

In the presence of oxygen the radiolysis of cysteine proceeds with a much higher yield suggesting the existence of a short-chain reaction [22]. Our experimental data have shown that the increasing yield in the presence of oxygen occurs also in GSH solutions irradiated at physiological pH as shown in Fig.1. It was also shown that almost all the GSH used up (up to 30% of the initial GSH) was converted to GSSG, as results from Fig.2. By analogy to what happens in the irradiation of cysteine, the following reactions can be postulated [22]:

(12)   GS• + O₂ ⟶ GSOO

(13)   GSOO + GSH ⟶ GSOOH + GS•

(8)   GS• + GS⁻ ⇌ GSSG⁻

(14)   GSSG⁻ + O₂ ⟶ GSSG + O₂⁻

(15)   O₂⁻ + GSH —H⁺→ GS• + H₂O₂

Reactions (12) and (14) have been shown by pulse radiolysis of cysteine [24]. They actually occur also in glutathione solutions and the relative rate constants, as resulting from our experiments, are:

$$k_{12 \, (GS + O_2)} = 1.6 \times 10^9 \cdot M^{-1} \cdot sec^{-1}$$
$$k_{14 \, (GSSG^- + O_2)} = 1.6 \times 10^8 \cdot M^{-1} \cdot sec^{-1}$$

At physiological pH, reaction (8) appears to effectively compete with reaction (12) making it likely that at this pH the propagating step is mainly reaction (14). And this is confirmed by the concomitant increase in the yield of GSSG and H_2O_2 at pH7.

Termination steps are likely to be reaction (9) and the following:

(16) $\quad GS^\cdot + GSOO \longrightarrow GSSG + O_2$

(17) $\quad GSOO + GSOO \longrightarrow GSSG + 2\, O_2$

The literature on steady-state radiolysis of oxidized glutathione is rather limited. Owen and Wilbraham [25] have reported that in air-saturated aqueous solutions of GSSG the sole product which can be identified is glutathione sulphonic acid, even with radiation doses destroying up to 80% of the starting material. On the basis of the available data and by analogy to the radiolysis of cysteine [26], the following series of reactions can be proposed:

(18) $\quad e^-_{aq} + GSSG \longrightarrow GSSG^-$

(19) $\quad GSSG^- \rightleftharpoons GS^\cdot + GS^-$

(20) $\quad H^\cdot + GSSG \longrightarrow GS^\cdot + GSH$

(21) $\quad OH^\cdot + GSSG \longrightarrow GS^\cdot + GSOH$

(22) $\quad GSOH + O_2^- \longrightarrow GSO_3^- + H^\cdot$

(23) $\quad GSOH + GSOH \longrightarrow GSO_2H + GSH$

(24) $\quad GSOH + H_2O_2 \longrightarrow GSO_2H + H_2O$

(12) $\quad GS^\cdot + O_2 \longrightarrow GSOO$

(25) $\quad GSOO + GSOO \longrightarrow GSO_2 + SO_2G$

(26) $\quad GSO_2^\cdot + SO_2G + H_2O \longrightarrow GSO_2H + GSO_3H$

The rate constants for reactions (18) and (21), as measured in our experiments, were:

$$k_{18 \, (e^-_{aq} + GSSG)} = 5 \times 10^9 \cdot M^{-1} \cdot sec^{-1} \text{ at pH7}$$
$$k_{21 \, (OH + GSSG)} = 9.9 \times 10^9 \cdot M^{-1} \cdot sec^{-1} \text{ at pH7}$$

Purdie [27] has reported that in the radiolysis of de-aerated solutions of some asymmetrical disulphides, the symmetrical disulphides were formed in very high yield suggesting the occurrence of chain reactions suppressed by oxygen. No reactions of this sort have been reported in the radiolysis of symmetrical disulphides and in particular of oxidized glutathione.

CONCLUSIONS

Considering at this point the mechanisms which have been proposed to account for the modification of radiosensitivity by thiol compounds, the following conclusions can be drawn.

The very high rate constants for the reaction of GSH and GSSG with water primary radicals make such compounds very well suited for the scavenging action. Moreover, glutathione is likely to have very good chances of being at the right place, at the right time to perform that action. Of course, being such action directed against radicals from water, its effect on radiation response of cells will only affect that part of damage due to the indirect effect of radiation, and the importance of the indirect effect of radiation in cellular radiobiology is still under discussion. Sanner and Pihl [28], for instance, have postulated that the indirect effect accounts for about 50% of radiation lethality in *Escherichia coli* B cells. There are authors, however, who believe that most of the biologically important radiation damage results from direct effect.

With regard to the mixed disulphide hypothesis, there is ample experimental evidence to show that glutathione and other protective SH compounds can form mixed disulphides with SH-proteins [29].

The model for the protective mechanism, as proposed by Eldjarn and Pihl [29], postulates that when the mixed disulphide bond is attacked by free radicals, one of the sulphur atoms becomes oxidized to a sulphinic or sulphonic acid, whereas the other one is reduced to an SH group. In this way the vital SH group has a 50% chance of escaping radiation inactivation. An analogous mechanism is postulated for the direct effect of radiation, implying interaction with water.

While theoretically possible, the mechanism is not supported by convincing experimental evidence, neither in vivo nor in model systems. In fact, model experiments with pure enzymes in solution have shown that the presence of mixed disulphides does not always really increase the radioresistance of the enzymes [30]. On the other hand, there is no real evidence so far to indicate that the inactivation of SH enzymes plays a significant role in cell killing by radiation.

The hydrogen donation model is probably the one that has received the better experimental support. The experiments and the interpretations proposed by Howard-Flanders and his co-workers [14,15], together with the experiments in simple model systems which provide direct evidence of the repair reactions [31,32], undoubtedly confer a particular credibility to this model.

The demonstration of the repair process given by the Adam's group [32] using the technique of pulse radiolysis is very convincing. By taking advantage of the strong and well-defined absorption of the radical anion $RSSR^-$, the authors have shown that in appropriate conditions the anion can be formed in the following reactions:

$$XH + OH \longrightarrow X^{\bullet} + H_2O$$

$$X^{\bullet} + RSH \longrightarrow XH + RS^{\bullet}$$

$$RS^{\bullet} + RS^- \rightleftharpoons RSSR^-$$

Such a mechanism has been shown to operate between cysteamine and hydrogen sulphide and between aliphatic alcohols, glucose and polyethylene oxide of various molecular weights [31,32]. The repair process is inhibited by oxygen and is pH dependent.

In conclusion, radiation chemical data indicate that mechanisms like those discussed here can operate in living cells and justify the influence of glutathione on radiation sensitivity. However, more biological data are still needed to sort out whether glutathione really plays any part in modulating radiation sensitivity, and, if so, which mechanism(s) is (are) involved.

One point which needs some further comments is that concerning the reasons why exogenous sulphydryls are generally more effective in protecting cell systems when irradiations are given in the presence of oxygen. No obvious experimentally supported explanation can be proposed at present.

It has to be pointed out that while the dose reduction factor (DRF) brought about by exogenous sulphydryls is much higher in the presence of oxygen than in its absence, the absolute radiosensitivity of cells protected in anoxia is always lower than that of cells protected by the same concentration of sulphydryl in the presence of oxygen. This may suggest the existence of a maximum value of the DRF corresponding, for instance, to the scavenging of all radicals responsible for the indirect effect of radiation or else to the restitution of all the damage susceptible to be repaired by hydrogen donation. It is reasonable to suppose, therefore, that in anoxic conditions the radiosensitivity in the absence of any exogenous protector is already fairly close to the minimum value of radiosensitivity which can be obtained. The opposite situation occurs in the presence of oxygen. The existence of chain reactions in the radiolysis of thiol compounds in the presence of oxygen, in which oxygen is used at the same rate as the thiol and converted into hydrogen peroxide or water, appears to be an effective mechanism competing with the fixation by oxygen of radiation damage to critical molecules or with the formation of harmful oxygen radicals. In this way the high values of DRF can be justified.

REFERENCES

[1] BACQ, Z.M., Chemical Protection Against Ionizing Radiation, Charles C. Thomas, Springfield, Illinois (1965).
[2] JOCELYN, P.C., Biochemistry of the SH Group, Academic Press, London, New York (1972).
[3] RÉVÉSZ, L., MODIG, H., Cysteamine-induced increase of cellular glutathione level: a new hypothesis of the radioprotective mechanism, Nature (London) 207 (1965) 430.
[4] CASPERSSON, O., REVESZ, L., Cytochemical measurement of protein sulphydryls in cell lines of different radiosensitivity, Nature (London) 199 (1963) 153.
[5] OHARA, H., TERASIMA, T., Variations of cellular sulphydryl content during cell cycle of HeLa cells and its correlation to cycling change of X-ray sensitivity, Exp. Cell Res. 58 (1969) 182.
[6] SINCLAIR, W.K., "Dependence of radiosensitivity upon cell age", Time and Dose Relationships in Radiation Biology as Applied to Radiotherapy (NCI-AEC Conference, Carmel, California), Brookhaven National Laboratory BNL-50203 (C-57) (1970) 97.
[7] SINCLAIR, W.K., Cell cycle dependence of the lethal radiation response in mammalian cells, Curr. Top. Radiat. Res. Quart. 7 (1972) 264.
[8] SINCLAIR, W.K., N-ethylmaleimide and the cyclic response to X-rays of synchronous Chinese hamster cells, Radiat. Res. 55 (1973) 41.
[9] HARRIS, J.W., PAINTER, R.B., HAHN, G.M., Endogenous non-protein sulphydryl and cellular radiosensitivity, Int. J. Radiat. Biol. 15 (1969) 289.
[10] HARRIS, J.W., POWER, J.A., Diamide: a new radiosensitizer for anoxic cells, Radiat. Res. 56 (1973) 97.
[11] VOS, O., BUDKE, L., VERGROESEN, A.J., Protection of tissue culture cells against ionizing radiation, I. The effect of biological amines, SS compounds and thiols, Int. J. Radiat. Biol. 5 (1963) 543.
[12] ELDJARN, L., BREMER, J., BORRESEN, H.C., The reduction of disulphides by human erythrocytes, Biochem. J. 82 (1962) 192.
[13] MODIG, H.G., EDGREN, M., REVESZ, L., Release of thiols from cellular mixed disulphides and its possible role in radiation protection, Int. J. Radiat. Biol. 22 (1971) 257.
[14] HOWARD-FLANDERS, P., LEVIN, J., THERIOT, L., Reactions of DNA with sulphydryl compounds in X-irradiated bacteriophage systems, Radiat. Res. 18 (1963) 593.
[15] JOHANSEN, I., HOWARD-FLANDERS, P., Macromolecular repair and free radical scavenging in the protection of bacteria against X-rays, Radiat. Res. 24 (1965) 184.
[16] HUTCHINSON, F., Sulphydryl groups and the oxygen effect on irradiated dilute solutions of enzymes and nucleic acids, Radiat. Res. 14 (1961) 721.
[17] KOHN, N.J., GUNTER, S.E., Cysteine protection against X-rays and the factor of oxygen tension, Radiat. Res. 13 (1960) 250.
[18] VERGROESEN, A.J., BUDKE, L., VOS, O., Protection of tissue culture cells against ionizing radiation. III. The influence of anoxia on the radioprotection of tissue culture cells by cysteamine, Int. J. Radiat. Biol. 6 (1963).

[19] PIHL, A., SANNER, T., "Chemical protection against ionizing radiation by sulphur-containing agents", Radiation Protection and Sensitization (MOROSON, H.L., QUINTILIANI, M., Eds) Taylor & Francis Ltd., London (1970) 43.
[20] ADAMS, G.E., Radiation chemical mechanisms in radiation biology, Adv. Radiat. Chem. **3** (1969) 125.
[21] BONOTTO, S., NETRAWALI, M.S., Radiolysis and radioprotective effect of glutathione, Int. J. Radiat. Biol. **15** (1969) 275.
[22] LAL, M., ARMSTRONG, D.A., WIESER, M., The cobalt-60 gamma-radiolysis of reduced glutathione in deaerated aqueous solutions, Radiat. Res. **37** (1969) 246.
[23] AL-THANNON, A.A., BARTON, J.P., PACKER, J.E., SIMS, R.J., TRUMBORE, C.N., WINCHESTER, R.V., The radiolysis of aqueous solutions of cysteine in the presence of oxygen, Int. J. Radiat. Phys. Chem. **6** (1974) 233.
[24] BARTON, J.P., PACKER, J.E., The radiolysis of oxygenated cysteine solutions at neutral pH. The role of RSSR$^-$ and O$_2^-$, Int. J. Radiat. Phys. Chem. **2** (1970) 159.
[25] OWEN, T.C., WILBRAHAM, A.C., Glutathionesulfonic acid, the predominant product of X-radiolysis of air-saturated solutions of oxidized glutathione, Radiat. Res. **50** (1972) 253.
[26] PURDIE, J.W., γ-radiolysis of cysteine in aqueous solution. Dose-rate effects and a proposed mechanism, J. Am. Chem. Soc. **89** (1967) 226.
[27] PURDIE, J.W., Investigation of chain reactions and oxygen effects during radiolysis of peptide disulphide bonds using cysteine-glutathione disulphide as a model, Radiat. Res. **48** (1971) 474.
[28] SANNER, T., PIHL, A., Significance and mechanism of the indirect effect in bacterial cells. The relative protective effect of added compounds in *Escherichia coli* B irradiated in liquid and in frozen suspensions, Radiat. Res. **37** (1969) 216.
[29] ELDJARN, L., PIHL, A., "Mechanisms of protective and sensitizing action", Ch. 4, Vol. 2, Mechanisms in Radiobiology (ERRERA, M., FORSSBERG, A., Eds), Academic Press, New York & London (1960).
[30] QUINTILIANI, M., BOCCACCI, M., Factors affecting the *in vitro* inactivation of aldolase by X-rays, Int. J. Radiat. Biol. **7** (1963) 255.
[31] ADAMS, G.E., McNAUGHTON, G.S., MICHAEL, B.D., Pulse radiolysis of sulphur compounds. Part 2. – Free radical "repair" by hydrogen transfer from sulphydryl compounds, Trans. Farad. Soc. **64** (1968) 902.
[32] ADAMS, G.E., ARMSTRONG, R.C., CHARLESBY, A., MICHAEL, B.D., WILLSON, R.L., Pulse radiolysis of sulphur compounds. Part 3. Repair by hydrogen transfer of a macromolecule irradiated in aqueous solution, Trans. Farad. Soc. **65** (1969) 732.

INHIBITION OF DNA REPAIR BY CHEMICAL AND BIOLOGICAL AGENTS*

H. ALTMANN, Helga TUSCHL, E. WAWRA,
Ingrin DOLEJS, W. KLEIN, A. WOTTAWA
Institut für Biologie,
Forschungszentrum Seibersdorf,
Seibersdorf, Austria

Abstract

INHIBITION OF DNA REPAIR BY CHEMICAL AND BIOLOGICAL AGENTS.
The possibility to radiosensitize tumour cells by repair inhibitors coupled to site-specific carriers is discussed. The effect of some drugs on DNA repair, normal DNA synthesis and on isolated enzymes involved in DNA repair was tested. Some preliminary experiments were performed with mycoplasmas to test the possibility of using biological factors in combination with radiation in the treatment of malignant tumours. Mycoplasma infection was shown to interfere with DNA repair and semi-conservative DNA synthesis in the spleen cells of rats.

INTRODUCTION

Bacterial systems are known to have good correlation between their capacity for DNA excision repair and the radioresistance of the cells. Inhibition of excision repair lowers the shoulder of survival curves because of the accumulation of premutation lesions. If the amount of damage in DNA exceeds the capacity of the excision repair, the unfilled gaps remain after replication in the daughter DNA-strand. If this does not cause cell death these gaps will either give rise to mutations or they will be targets of post-replication repair, but mutations may also occur as a result of this error-prone repair process.

The mechanisms just mentioned are important for the inactivation of microorganisms but are only of little value in tumour radiotherapy because the selectivity of DNA repair inhibitors directed only to tumour cells and not to other body cells is rather low.

Treatment with radiosensitizers directed to the site of the cancer is one approach in modern cancer therapy.

Three ways to improve this approach are commonly under investigation:

(a) The use of radiosensitizers specific to hypoxic cells.
(b) Coupling of DNA repair inhibitors or other chemical compounds with tumour-specific antibodies.
(c) Enrichment of cytotoxic microorganisms in tumours.

The first aspect is discussed extensively in other papers in these Panel Proceedings.

Problem (b) is relatively new but it has positive aspects especially because methods have improved for the isolation of tumour-specific antibodies directed against specific surface antigens. One difficulty is the preparation of complexes which preserve both the DNA repair inhibition capacity and the ability of antibodies to recognize the tumour antigens. Degradation of the carrier protein in the area of fast tumour growth by lysosomal proteases is important for this proposed mechanism to take place.

---

* Supported in part by a grant from the Austrian "Bundesministerium für Gesundheit und Umweltschutz".

Besides protein carriers, liposomes may also be used as carriers. In future, the entrapment of repair inhibitors into experimentally produced liposomes followed by degradation with the aid of lysosomal lipases near the target cells should also be taken into consideration.

DNA excision repair is mainly responsible for the genetic integrity. A normal excision repair keeps the natural mutation rate at a constant level. If excision repair is inhibited, post-replication repair comes into action like a fire brigade when the survival of the cells is endangered. But this post-replication repair is often error prone resulting in an increase in the mutation rate.

As shown in mouse lymphoma cells the gap-filling post-replication repair is largely dependent on DNA "de novo" synthesis [1]. This recombinational repair in eukaryotic cells is possibly even more error prone than in prokaryotes because of the repetitive sequences in their DNA.

Therefore, to investigate the sensitization of tumour cells for radiation therapy it might be better to use inhibitors of post-replication repair for sensitizing the cells instead of inhibitors of DNA excision repair. Post-replication repair is dependent on the "de novo" synthesis of a section of DNA and can possibly be inhibited by inhibitors of semiconservative DNA synthesis.

The following reasons seem to render this concept practicable:

1. DNA excision repair is responsible for the genetic integrity of cells and is therefore an important resistance factor against diverse influences. Uncoupled inhibitors of DNA excision repair are tumour specific to a limited extent only and are therefore also able to influence other body cells resulting in a decrease of resistance of those cells to carcinogenic agents. This kind of action includes an increase in the risk of secondary tumours.
2. Inhibitors of semiconservative DNA synthesis strongly influence post-replication repair. This may help to lower the risk of metastases or secondary tumours being formed.

However, all these effects are not as simple as they seem at first sight and there are many other parameters that have to be taken into account.

All chemicals tested so far do not specifically influence only DNA repair but have also an inhibitory effect on semi-conservative DNA synthesis or on the synthesis of some other macromolecules. The effects of some drugs are strongly dependent upon concentration. At a certain concentration the influence of a chemical is considerable on normal DNA synthesis and less on DNA repair, and vice versa.

Section 1 discusses the mechanism by which some drugs and chemical agents interfere with DNA metabolism [2]. Section 2 shows that there are possibly other ways of treating cancer by using biological factors for radiosensitizing tumour cells.

TABLE I. INFLUENCE OF DRUGS ON DNA REPAIR IN MOUSE SPLEEN CELLS MEASURED EITHER BY THE KINETICS OF THYMIDINE INCORPORATION INTO DNA OR BY THE REJOINING OF STRAND BREAKS BY DENSITY GRADIENT CENTRIFUGATION IN ALKALINE SUCROSE

Drug (metabolized)	Concentration (μg/ml)	Per cent of control by:	
		thymidine incorporation	density gradient centrifugation
Vinblastine	2	102	100
Vincristine	2.5	106	60
Cyclophosphamide	50	79	70
Procarbazine	100	81	40

1. THE USE OF CHEMICAL COMPOUNDS AS INHIBITORS OF DNA REPAIR

In our initial research we tested some drugs already used in tumour therapy for their action on DNA repair.

Table I shows the influence of four different drugs on DNA excision repair measured with two different methods. In the first method we used thymidine incorporation into mouse spleen cells in the presence of hydroxyurea (10^{-2}M), and in the second method we investigated the rejoining of strand breaks in the DNA molecule by means of density gradient centrifugation in alkaline sucrose [3]. In both methods the DNA was damaged by irradiating the cells with a 30-krad ^{60}Co gamma source. All investigations were carried out with metabolized drugs and incubation times of 30 min at 37°C.

It seems of interest that the drug Vincristine shows no effect on DNA repair when tested by the thymidine incorporation method but with density gradient centrifugation there was an effect. These different effects could be explained by an inhibitory action of the drug on the ligase reaction.

Cyclophosphamide gave nearly the same inhibition values in both experiments whereas the results with Procarbazine indicate that a ligase inhibition was also at least partly responsible for the lower values obtained with density gradient centrifugation.

Table II shows the results of experiments on DNA repair, isolated DNase and DNA-dependent DNA polymerases as well as the results on semi-conservative DNA synthesis [2]. Estimations of semi-conservative and DNA repair syntheses were made by measuring thymidine incorporation in the DNA of mouse spleen cells. The enzyme experiments were performed in cell-free systems with DNA polymerase-α (large) and DNA polymerase-β (small) from pig spleen as well as with DNase-I from beef pancreas. Six of the drugs tested showed an inhibitory effect on semi-conservative DNA synthesis or DNA repair but no effect on DNA polymerase or DNase activity. A possible explanation of these results could be that these substances are not acting directly on repair enzymes but have considerable influence on the precursor pool.

A rather general inhibitory effect could be obtained with Nalidixic acid. Ethidium bromide showed very strong inhibitory effects but they seemed to be less specific possibly because of the intercalation reaction of this substance with DNA.

We also tested Tween 80 [4], an unspecific DNA repair inhibitor, by means of autoradiographic methods. Figure 1 shows that Tween 80 has a much more pronounced effect on the percentage of labelled cells than on the number of grains per labelled cell.

A new target for inhibition in tumour therapy is now under discussion, namely reverse transcriptase. Inhibitors of reverse transcriptase have little effect on the reduction of radio-resistance of cells but they could become effective immediately after radiotherapy.

Reverse transcriptase is concerned with only the initial step of cell transformation by RNA viruses and is not necessary for the maintenance of the transformed state.

Most of the inhibitors of reverse transcriptase that have been tested so far do not act specifically on this enzyme but have, more or less, the ability to inhibit DNA-dependent DNA-polymerase.

2. BIOLOGICAL FACTORS

Besides chemical treatment, another possible approach to tumour therapy might be the development of new methods using biological factors such as mycoplasmas for radiosensitizing tumour cells [5]. Mycoplasmas are frequently detected in malignant tumours as contaminants [6,7]. In addition, in-vitro experiments have shown that different strains of mycoplasmas have oncolytic effects on carcinoma cells [8]. The injection of mycoplasmas into solid 8-day-old carcinomas resulted in tumour destruction. The reaction mechanisms may be similar to those of viral oncolyses or other adjuvant effects including macrophage activation.

TABLE II. INFLUENCE OF SOME DRUGS ON CERTAIN STEPS OF DNA METABOLISM

For repair experiments cells were irradiated with 30 krad of ^{60}Co

Drug	Concentration (μg/ml)	Semi-conservative DNA synthesis	DNA repair after gamma irradiation	DNA polymerase Small	DNA polymerase Large	DNase
Vinblastine	2.0	++	–	–	–	–
Vincristine	2.5	–	–	–	–	–
Cyclophosphamide	50	–	+(M)	–	–	+(M)
Procarbazine	100	++	+(M)	–	–	–
Penicillin-G/ procaine-penicillin-G	750	–	++	–	–	–
Phenylbutazone	100	++	+++	–	–	–
Isoniazid	100	++	–	–	–	–
Nalidixic acid	100	+	++	+(M)	+(M)	+/++(M)
Ethidium bromide	100	+++	+++	+++	+++	+++

+ = slight inhibition, ++ = strong inhibition, +++ = very strong inhibition, – = no effect, (M) = metabolized drug

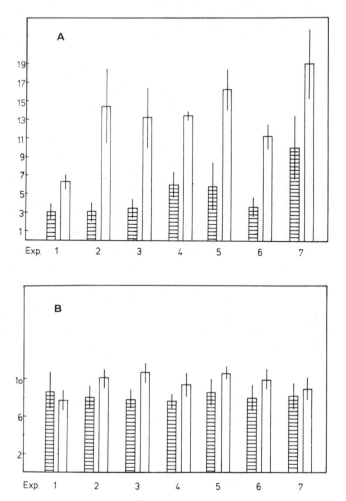

FIG.1. Autoradiographic estimation of DNA repair in mouse spleen cells after ^{60}Co irradiation and incubation with ^{3}H-thymidine in the presence of hydroxyurea. A = percentage of labelled cells, B = average number of grains per labelled cell (▤ 0.002% Tween 80, □ control).

In our preliminary experiments, we did not use tumour-bearing rats but injected mycoplasma suspensions (*Mycoplasma arthritidis*) into the peritoneal cavity of Sprague-Dawly rats [9,10]. After five days the rats developed polyarthritis.

During the acute and sub-acute stages of this mycoplasma-induced polyarthritis, experiments were performed to estimate semi-conservative DNA synthesis and DNA excision repair. Figure 2 shows that the rate of semi-conservative DNA synthesis (curve A) and the rate of unscheduled DNA synthesis (approximated by the region between curves C and D) are lower in rats during the acute stage of infection than in healthy control animals.

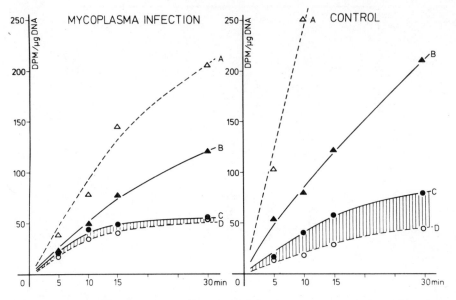

*FIG.2. Inhibition of DNA repair and semi-conservative DNA synthesis in the acute stage of mycoplasma infection (*Mycoplasma arthritidis, *spleen cells of Sprague-Dawley rats).*

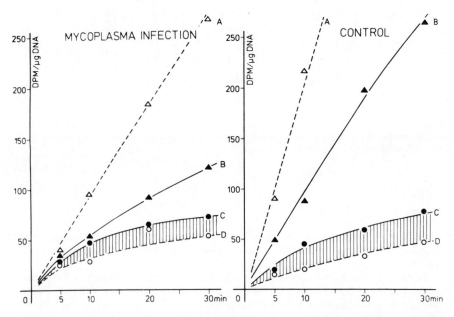

FIG.3. Nearly complete normalization of DNA repair and DNA synthesis in rat spleen cells three months after infection.

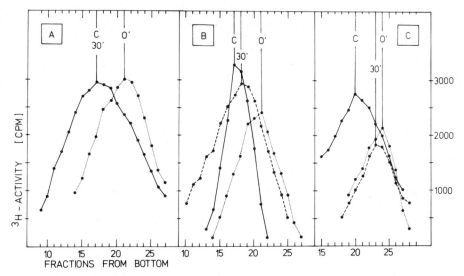

FIG.4. *Comparison of DNA strand rejoining after ^{60}Co irradiation (30 krad) of rat spleen cells by means of density gradient centrifugation. A = control, B = adjuvant arthritis (Freund's adjuvant), C = mycoplasma arthritis.*

In contrast, Fig. 3 indicates the almost total normalization of semi-conservative as well as unscheduled DNA synthesis when the animals were investigated during the sub-acute stage of the arthritis (12 weeks after infection). The DNA synthesis and the DNA repair are only slightly affected [11]. The experiments were done on spleen cells of rats by incorporating labelled thymidine into acid-insoluble material under appropriate conditions.

Similar experiments were performed using the method of density gradient centrifugation [12]. The DNA of the spleen cells of the animals was prelabelled with ^{3}H-thymidine and then the DNA was damaged by irradiating the cells with a 30-krad ^{60}Co gamma source. After the cells had been incubated for 30 min at 37°C the rejoining of the single-strand breaks was estimated by gradient centrifugation in alkaline sucrose.

The results of a typical experiment are shown in Fig. 4. In the cells of healthy animals the sedimentation profile of the DNA returns to the original position in the gradient within 30 min after the DNA has been damaged by ionizing radiation (A). If an arthritis is produced by applying Freund's adjuvant, the DNA of the animals shows nearly the same behaviour as that of healthy animals (B), whereas the spleen cells of rats in the acute stage of mycoplasma-induced arthritis almost completely lose their ability to rejoin single-strand breaks in their DNA (C).

Further experiments are being made with animal tumour models and different mycoplasma strains to examine the possibility of radiosensitizing tumours with these biological agents. Experiments with chemical inhibitors of semi-conservative DNA synthesis and/or of DNA repair and inhibitors of specific enzymes are also being carried out to investigate the possibility of their coupling to site-specific carriers.

REFERENCES

[1] LEHMANN, A.R., Postreplication repair of DNA in ultraviolet-irradiated mammalian cells, J. Mol. Biol. **66** (1972) 319.
[2] WAWRA, E., KLEIN, W., KOCSIS, F., WENIGER, P., Action of Some Drugs on Enzymes Involved in DNA Repair and Semi-conservative DNA Synthesis, SGAE-2469, BL-144/75.

[3] KLEIN, W., KOCSIS, F., ALTMANN, H., Die Wirkung einiger Zytostatika auf die DNS-Synthese von normalen und Tumor-Zellen, Arzneim.-Forsch. (Drug Res.) **25** (1975) 1623.

[4] TUSCHL, H., KLEIN, W., KOCSIS, F., BERNAT, E., ALTMANN, H., Investigations into the inhibition of DNA repair processes by detergents, Environ. Physiol. Biochem. **5** (1975) 84.

[5] GERICKE, D., Bösartige Geschwülste und Mikroorganismen, Fortschritte der Medizin **89** 1 (1971) 32.

[6] GERLACH, F., Tumoren als Spätfolge einer pränatalen Infektion mit Mykoplasman, Wien. Tierärztl. Monatsschr. **57** (1970) 232.

[7] GERICKE, D., SCHÜTZE, E., Personal communication.

[8] SETHI, K.K., BRANDIS, H., Oncolytic effect of a *Mycoplasma gallisepticum* strain on solid Ehrlich carcinoma, Pathol. Microbiol. **37** (1971) 105.

[9] KLEIN, G., WOTTAWA, A., STREIT, S., ALTMANN, H., DNA Repair and DNA Antibodies During Experimental Mycoplasma Arthritis, SGAE-2360, BL-116/74.

[10] LABER, G., SCHÜTZE, E., TEHERANI, D., TUSCHL, H., ALTMANN, H., Correlation Between the Occurrence of Mycoplasma and DNA Repair Inhibition in Spleen Cells of Rats During Experimental Mycoplasma Arthritis, SGAE-2364, BL-120/74.

[11] KLEIN, G., Enzymuntersuchungen zur Pathogenese der experimentellen Mykoplasma-Arthritis, Z. Rheumatol. **34** (1975) 162.

[12] GEREICKE, D., SCHÜTZE, E., Mycoplasmen in Transplantationstumoren kleiner Versuchstiere, Zentralbl. Bakteriol. Orig. I. **208** (1968) 329.

SYNERGISTIC EFFECT OF RADIOPROTECTIVE SUBSTANCES HAVING DIFFERENT MECHANISMS OF ACTION

L.B. SZTANYIK, A. SÁNTHA
"Fréderic Joliot-Curie" National
Research Institute for Radiobiology and Radiohygiene,
Budapest, Hungary

Abstract

SYNERGISTIC EFFECT OF RADIOPROTECTIVE SUBSTANCES HAVING DIFFERENT MECHANISMS OF ACTION.
Twenty years of research on chemical protection against the harmful effects of ionizing radiation are reviewed. Research activity has been developed in three main directions. Alterations in the basic molecular structure of one of the well-established radioprotective substances, AET, have not resulted in less toxic and more efficient derivatives than the original compound. Pharmacodynamic effects of AET have been counterbalanced by corresponding pharmacological antagonists, but the expected decrease in general toxicity has not been achieved. Combinations of radioprotective substances, such as AET and cysteine, AET and methoxytryptamine, AET and mercaptopropionylglycine have led to the most significant improvement in protective activity and diminution in toxicity. Combinations of radioprotectors having different mechanisms of action might deserve further attention.

INTRODUCTION

More than a quarter of a century has passed since the protection of animals against the acute lethal effect of radiation by chemical means was first demonstrated in 1949. An enthusiastic research activity started immediately after this discovery in many laboratories all over the world.

The intensive research, however, has failed to produce any clinically applicable radioprotective drug. Compounds, which in animal experiments appeared to be the most effective and promising, proved to be extremely toxic for humans. This is one of the reasons why interest in and enthusiasm for the studies on chemical radiation protection have declined sharply during the last decade. All the more reason why, in the meantime, efforts in classical pharmacological research have resulted in numerous excellent medicaments that are being used in clinical practice with considerable success, such as psycho-pharmaka, anti-tuberculotics, beta-sympatholytics, cytostatics used as anti-tumour drugs, oral diuretics and anti-diabetics, and last but not least oral contraceptives.

Our own research in the field of chemical radiation protection was initiated in 1957 and developed in three directions:

1. Synthesis and testing of new, potentially radioprotective substances derived from well-established ones by systematic alterations in the molecular structure;

2. Application of pharmacological antagonists to eliminate toxic side-effects of otherwise effective radioprotectors;

3. Combination of radioprotective agents having different mechanisms of action to decrease toxicity and increase radioprotective efficiency.

The present paper is a brief account of our almost 20 years' work in chemical radioprotection.

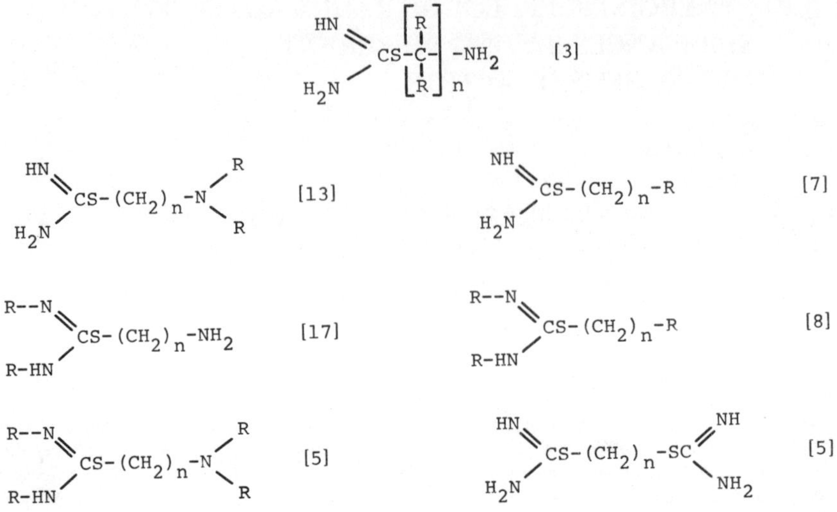

FIG.1. Chemical structure of AET derivatives tested.

1. ALTERATIONS IN THE CHEMICAL STRUCTURE OF AET

The studies reported here were started soon after the first papers on the radioprotective effect of AET (S, 2-aminoethylisothiuronium-Br-HBr) were published by Doherty et al. [1–3]. Stability, solubility and, in particular, radioprotective efficiency of this compound seemed to be superior to those of other radioprotective substances known at that time: cysteine, cysteamine, cystamine, glutathione, serotonin, etc. Therefore, we decided to concentrate our research efforts on a detailed analysis of AET.

Parallel to the studies on toxicity, metabolism, and pharmacological and radioprotective activity of AET, almost a hundred different derivatives were also synthesized and tested in 1957–72. The latter investigations were made to explore the importance of various structural elements of the AET molecule in its biological properties. The general structure of these compounds is given in Fig. 1.

On the basis of our own experimental results and data reported in the literature, the structural and functional inter-relationships in the AET molecule can be summarized as follows [4–13]:

(a) The hydrocarbon chain should not be made up of more than three carbon atoms. The radioprotective effects of the ethyl and propyl homologues do not differ significantly. Symmetric substitution of two hydrogen atoms in the carbon chain of AET has resulted in an equivalent decrease of toxicity and radioprotective activity. In contrast to this, an asymmetric substitution has produced a compound with good radioprotective effect, but very high toxicity (2-ABT).

(b) Toxicity of simple alkyl-isothiouroniums, containing no amino group at all, has been about half that of the corresponding aminoalkyl derivatives. The first three members of the homologous line have exerted moderate radioprotective effects but higher homologues have been ineffective.

(c) Substitution of one or both hydrogens of the amino group leads to an increase in toxicity and a decrease in radioprotective effect.

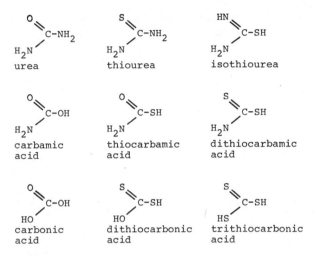

FIG.2. Chemical structure of isothiourea-related compounds studied.

Derivatives, in which the amino group has been enclosed into a heterocyclic ring, such as piperidine, retained radioprotective activity. It should be noted here that similar compounds, containing a heterocyclic ring instead of the amino group, have been found totally ineffective by Shapira et al. [3].

(d) Substitutions in the isothiouronium group of AET have decreased both the toxicity and radioprotective activity to about the same degree. Of these compounds, the N'-methyl derivative of AET proved to be the most effective. Doherty et al. have found N'-methyl and N'-dimethyl derivatives of APT less toxic than the original compound but equally radioprotective.

Enclosure of the isothiouronium group into a ring of 5 or 6 atoms is accompanied with an increase in toxicity. Imidazole analogues are 1.5–3 times and tetrahydropyrimidine analogues are 2.5–5 times more toxic than the corresponding isothiouroniums. Imidazole analogues of AET and APT have a moderate radioprotective effect on mice and lower organisms; tetrahydropyrimidines have only an effect on single cells.

(e) Cyclic analogues of alkyl-isothiouroniums do not have a radioprotective effect on either mice or single cells.

(f) Even the higher homologues of bis-alkylene-isothiouronium compounds are lacking radioprotective activity, although these substances are pharmacologically active [14]. Consequently, the presence of an isothiouronium group in a molecule per se is not sufficient for the radioprotective effect.

(g) Increasing the number of amino groups in the molecule has not led to increasing the radioprotective effect. Similarly, replacement of the guanidine group in the transformed molecule by a more basic structural element, for instance hydrazine, has also remained ineffective.

In addition to the foregoing "direct derivatives of AET", several other compounds in which isothiourea has been replaced by other molecular structures resembling isothiourea were also synthesized and tested. It had been known before that compounds which contained no free thiol groups, but were capable of transforming into thiol compounds in the organism, were in general more stable, less toxic and still radioprotective. Such compounds are the esters of amino-alkylthiols with aliphatic acids, amino acids, sulphuric acid or phosphoric acid. The general structure of the compounds studied by us can be seen in Fig. 2.

Aminoalkyl-dithiocarbamates deserve particular attention as they are rather stable, less toxic and almost as effective as MEA and AET. They are structurally closely related to AET and essentially consist of two independently radioprotective components, mercapto-alkylamine and dithiocarbamic acid. The ammonium salt of the latter has been found by van Bekkum [15] to be as effective as cysteine as a radioprotector.

Concerning the order of radioprotective effectivity, dithiocarbamates are followed by alkyl- and aminoalkyl-trithiocarbonates and the esters of trithiocarbonic acid with adequate alkyl- or aminoalkyl-thiols. These are also less toxic than the corresponding isothiouronium derivatives, but they are rather unstable. Too many sulphur atoms on the same carbon atom seems to weaken the stability of the molecule.

Esters of aminoalkylthiols with carbamic acid and alkylthiols with dithiocarbonic acid do not have convincing radioprotective effects. The common characteristic feature of these substances is that their molecules contain oxygen atoms. Accordingly, if the sulphur atom is replaced by the chemically related oxygen atom, the radioprotective activity is influenced unfavourably.

On the basis of the previously reviewed experimental results, it can be concluded that any essential change in the original structure of aminoalkylisothiouronium compounds (AET and APT) such as substitution of hydrogen atoms in the carbon chain, replacement of amino- or isothiouronium groups by other functional groups, or enclosure of these groups into cyclic structures, has not resulted in any substances with more favourable biological properties, in particular with lower toxicity and better radioprotective efficiency, than the original AET.

2. COMBINATION OF AET WITH ITS PHARMACOLOGICAL ANTAGONISTS

The pharmacological properties of AET were first described by DiStefano et al. [16] in 1956 and further studied by several other researchers in the subsequent years [17, 18].

These studies have revealed that small doses of AET bring about a transient fall in arterial blood pressure, bradycardia and apnea. The characteristic triplet of symptoms is attributed to chemical stimulation of vagus receptors, i.e. to a cholinergic effect which can be prevented by vagotomy or parasympatholytic drugs such as atropine and scopolamine. With larger doses or repeated administration of AET, a more pronounced and prolonged hypotension is induced which cannot be influenced with either parasympatholytics or with antihistamine drugs. Pharmacological investigations have proved that this phenomenon is a ganglionic blocking effect. Sometimes, a transient increase in blood pressure is observed as a consequence of a direct stimulatory effect on the walls of the blood vessels. This can be diminished by vasodilatating agents of peripheral action, for instance Arfonad (Trimetaphan). In extreme large doses, AET causes convulsions, paralysis of the respiratory centre, and death.

Administration of AET to animals is often followed by salivation, nausea and vomiting induced by direct excitation of the emetic centre or via a reflex mechanism.

It has been established that these untoward pharmacological effects of AET do not play any role in its radioprotective activity, but they constitute a serious limitation to its clinical application [7]. That is why a series of experiments was carried out by us to find pharmacological antagonists that might be suitable to counterbalance these undesired side-effects.

Because of the high parasympathomimetic effect of AET, atropine was tested first as an antagonist. Bradycardia induced by AET in a dose of 280 mg/kg or more, which is sufficient to provide a noticeable radioprotective effect, was successfully compensated with a high dose of atropine (50 mg/kg), while other toxic effects of AET remained mostly unaffected. Sympathomimet drugs, such as adrenaline, noradrenaline and ephedrine, did not influence favourably the cardiotoxic effect of AET, except in extremly large doses: 10–20 mg/kg. Promethazine and diazepam, representatives of major and minor tranquillizers, administered together with AET, could counterbalance the increased "tone" of the vagus nerve only temporarily. Neither of these drugs prevented extrasystoles caused by large doses of AET.

Since the favourable effect of caffeine on cardiac and respiratory functions has been known for a long time, it was also tested by us in combination with AET. Doses of 1 to 5 mg/kg of caffeine decreased vagotonia markedly, although other disturbances persisted.

According to the literature [19], small doses of the excellent antiemetic drug metochlopramide (methoxychloroprocaineamide, Primperan[R] — Delagrange, Paris) acts favourably on cardiac function. Our experiments indeed proved that no disturbance in cardiac rhythm was induced by AET in the presence of metochlopramide. The drug could also prevent the emetic effect of AET on pigeons which are particularly sensitive to this compound.

Spiractin (1-piperidino-methylcyclohexanone), a lobeline-like drug used to stimulate the respiratory centre, was able to reduce the suppression of breathing activity induced by AET.

Results obtained with combinations of caffeine, metochlopramide, spiractin and aparkazin, a member of the chlorpromazine-type drugs used for treatment of parkinsonism, have shown that some of the toxic symptoms of AET can be satisfactorily abolished if it is given simultaneously with these drugs [20, 21]. We have supposed, therefore, that the addition of one or more pharmacological antagonists to AET can probably increase the tolerance of animals to AET and cancel the present limits of its clinical use. However, these expectations were not fulfilled in our later experiments on the joint toxicity of AET and its pharmacological antagonists.

When atropine, metochlopramide, spiractin and DMPP (1, 1-dimethyl-4-phenyl-piperazinium iodide), a powerful ganglion-stimulating agent, are each given to mice in sub-toxic doses either before or simultaneously with AET, the toxicity of AET is enhanced considerably. The decrease of LD_{50} of AET in the presence of atropine, metochlopramide or spiractin was between 15 and 40%. A more dramatic decrease of LD_{50} of AET, down to one seventh of its normal value, was obtained after the administration of DMPP [22].

In conclusion, we may note that some of the toxic phenomena induced by AET in animals can easily be counterbalanced by administration of proper pharmacological antagonists, but the general toxicity of the compound can hardly be influenced in this way.

3. COMBINATIONS OF AET WITH OTHER RADIOPROTECTORS

The failure of research efforts to obtain more powerful and less toxic radioprotectors among aminothiol and indolylamine compounds has prompted scientists to test combinations of radioprotective substances having different modes of action. Several papers were published already in the early 1960s describing the advantages of joint administration of two or more radioprotective agents [23–25]. The number of such publications has increased rapidly during the subsequent years.

Some of the results of our studies on the synergistic effect of combinations of AET and 5-methoxytryptamine (MOT), AET and cysteine (CSH), and AET and 2-mercaptopropionyl-glycine (MPG) are presented briefly here. Details of the investigations have been reported elsewhere [26–28].

3.1. Combination of AET with MOT

Not long after the radioprotective effect of serotonin (5-hydroxytryptamine) was discovered by Gray et al. [29] and confirmed by other scientists, derivatives of this indolylalkylamine compound were also prepared and tested for radioprotective activity. Of these derivatives, MOT has proved to be the most promising [30–32].

Although the toxicity of MOT is about twice that of AET, their radioprotective effectiveness does not differ significantly. Under the conditions of our experiments, the LD_{50} of AET given intraperitoneally to mice proved to be 580 ± 35 mg/kg (somewhat more than 2 mM/kg) and that of MOT was 190 ± 15 mg/kg (i.e. about 1 mM/kg). AET and MOT improve the survival of irradiated mice in proportion to the doses injected. The optimum protective dose of AET, 1 mM/kg, has a slightly greater dose-reduction factor (DRF) than 1/4 mM/kg of MOT (1.65 versus 1.60). The

DRF in these experiments has been calculated as the ratio of radiation doses producing 50% mortality within 30 days in the presence and in the absence of the radioprotector tested.

A pharmacological method used for studying the synergism of drugs was adopted by us for quantitative analysis of the joint toxicity of AET and MOT, and other compound combinations [26].

According to this method, doses of one component are recorded on the abscissa and those of the other component on the ordinate of a co-ordinate system in proportion to their own 50% lethal dose. Should the joint toxicity be strictly additive, possible combinations of the two compounds inducing an identical biological effect (in this case 50% mortality) will be represented by the diagonal line, or rather its 95% confidence limit range, connecting the corresponding dose-values of the individual components. A synergistic effect of two compounds is called potentiation if the combination producing a 50% effect is made up of doses less than those represented by the diagonal line. These points fall below the line. On the other hand, combinations located above the diagonal line express drug antagonism.

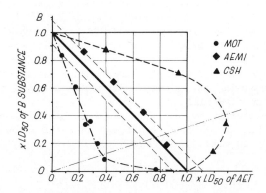

FIG.3. *Synergism of AET and other radioprotective substances in toxicity. AEMI (2, β-aminoethylmercapto-imidazoline) is an AET derivative in which the amidine group is enclosed into an imidazole ring.*

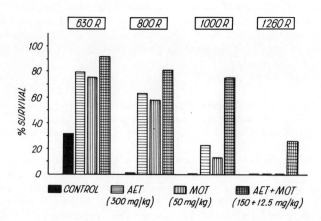

FIG.4. *Survival of mice exposed to 630–1260 R of X-rays and pretreated with AET, MOT and their combination.*

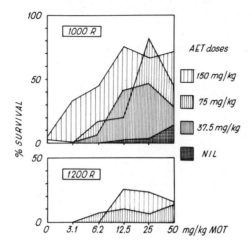

FIG.5. *Radioprotective effect of various combinations of AET and MOT on mice exposed to 1000 and 1200 R of X-rays.*

As can be seen in Fig. 3, synergism of AET and MOT in toxicity suggests a definite potentiation. Moreover, the isobole is apparently asymmetric indicating that the two compounds do not play an equal role in enhancing toxicity. The maximum of the curve is defined by the co-ordinates X = 0.35 and Y = 0.125. Accordingly, 50% mortality of mice can be achieved with the smallest doses of AET and MOT when the combination contains the individual constituents in a ratio of about 5 to 1 on a molar basis.

Combinations of these radioprotectors in adequate proportions increase the survival of mice irradiated with lethal doses in the range of 630 to 1260 R more efficiently than each of the components given alone in larger doses (Fig. 4). The $LD_{50/30}$ of mice treated with 150 mg/kg of AET and 12.5 mg/kg of MOT was increased from 500 ± 30 R to 1025 R which corresponds to a DRF of 2.05. Obviously, this combination has resulted in a survival that cannot be achieved by either AET or MOT given separately.

The radioprotective efficiency of AET and MOT combinations was studied in detail on mice exposed to 1000 R of X-rays. The optimum amount of AET given alone resulted in a 22.8% survival and that of MOT in a 13.3% survival only. By combining these two compounds in various proportions, the survival of mice has been increased far more than would have resulted from simply adding the effects of AET and MOT. The combination containing 75 to 150 mg/kg AET and 12.5 to 25 mg/kg MOT proved to be the most effective giving 45–75% survival. The graph suggests a co-operative effect of the two compounds in their protective activity that might be more than a simple addition (Fig. 5).

A quantitative analysis of the radioprotective effect has indicated that the synergism is of potentiating character. Combinations of AET and MOT in the proportions just mentioned are indeed the most effective.

3.2. Combination of AET with CSH

The sulphur-containing amino acid, cysteine, plays an important role in the structure and function of many common proteins. In pharmacology, cysteine is known as an antidote used in heavy-metal intoxications (cobalt and copper) and an antagonist of histamine. It was one of the

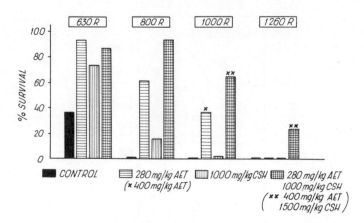

FIG.6. Survival of mice exposed to 630–1260 R of X-rays and pretreated with AET, CSH and their combination.

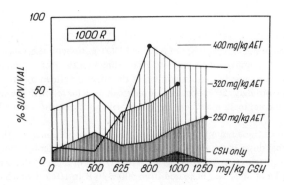

FIG.7. Radioprotective effect of various combinations of AET and CSH on mice exposed to 1000 R of X-rays.

very first chemicals shown to exert some protection against the acute effects of radiation including the death of mammals. According to Patt et al. [33], about 40% dose-reduction can be provided by 1200 mg/kg of cysteine (i.e. almost 10 mM/kg) in the dose range of 100 to 1000 R.

In our experiments, the LD_{50} of cysteine hydrochloride, injected intraperitoneally into mice, was found to be 2370 ± 300 mg/kg. The joint toxicity of AET and CSH in combination has suggested some antagonism between the two compounds, particularly in certain regions (see Fig. 3). The isobole of toxicity is evidently asymmetric, indicating that only CSH diminishes the toxic effect of the other component and that AET does not. The maximum of the curve is defined by the co-ordinates X = 1.3 and Y = 0.40. These co-ordinates suggest that an optimum combination, which has the least toxic effect on animals, contains AET and CSH in the ratio of 1 to 2.5 on a molar basis.

A mixture of AET and CSH can improve the survival of mice irradiated with doses between 630 and 1260 R more effectively than can each agent given alone. The optimum dose of CSH (1000 mg/kg) has a DRF of 1.35, and that of AET (280–400 mg/kg) has 1.65. The $LD_{50/30}$ of mice pretreated with a combination of 240–400 mg/kg AET and 1000–1500 mg/kg CSH has been increased from 500 ± 30 R to 1075 R, which corresponds to a DRF of 2.15 (Fig. 6). Accordingly, proper combinations of AET and CSH result in a significant decrease of toxicity and a considerable improvement in radioprotective activity, as has already been established by a number of scientists.

The protective effect of combinations of AET and CSH in various proportions has been studied in detail in mice exposed to 1000 R of X-rays (Fig. 7). The optimum amount of CSH given alone before the animals were irradiated resulted in a mere 7% survival, whereas that of AET resulted in 35%. The combination of 500–1250 mg/kg of CSH with 250–400 mg/kg of AET increased significantly the survival of irradiated mice. However, neither these results nor the DRF values mentioned before can provide any information on the character and degree of co-operation of these substances in radioprotection.

Quantitative analysis of synergism in radioprotective activity is much more complicated than that of toxicity. Very often the protective substances do not result in 50% survival of supralethally irradiated animals, if given alone. Nevertheless, this difficulty can be overcome for AET by extrapolating survival data obtained with doses of the compound less than the maximum tolerated. In such a way, the dose of AET needed to provide 50% survival of mice irradiated with 1000 R appeared to be about 560 mg/kg, i.e. very close to its LD_{50} toxic value. If an identical radioprotection could be achieved with about 2000 mg/kg of CSH, their synergism in radioprotective activity would be additive. However, a potentiation has to be assumed if doses of CSH larger than that are necessary (theoretically) to provide 50% survival [27].

It might also be mentioned here that a combination of radioprotectors made up of 150 mg/kg of AET, 12.5 mg/kg of MOT and 500 mg/kg of CSH did not give a better survival of mice exposed to 1000 or 1260 R of X-rays than did a mixture of the same amounts of AET and MOT [28].

3.3. Combination of AET with MPG

Sugahara et al. [34] reported in 1970 that Thiola^R, or MPG (2-mercaptopropionylglycine), exerts a moderate but consistent radioprotective effect if given before irradiation. Thiola is a relatively new detoxicating agent and thiol-source developed by Santen Pharmaceutical Co. in Japan. Its remarkably low toxicity urged one of the authors to study the possibility of applying this compound in combination with other radioprotective agents [35].

An aqueous solution of the original substance is acidic (pH 2.8) which necessitates neutralization before administration to animals. In a saline solution neutralized with sodium hydroxide, the LD_{50} of the compound for mice has been found to be approximately 3000 mg/kg (more than 15 mM/kg) in intraperitoneal injection, and above 10 000 mg/kg per os. A dose of 1600 mg/kg of MPG has resulted in 50% mortality of rabbits injected intravenously. The LD_{50} of a Thiola injection (Farmaceutici, Milan, Italy), a solution in distilled water and neutralized with sodium hydroxide, was found to be 1300 mg/kg for mice if given intraperitoneally.

According to our physico-chemical investigations, an explanation for these conflicting data might be that the substance, under certain conditions, tends to undergo transformation into its dimer, propionylglycine disulphide (PGD), which is less toxic than the original monomer. (This PGD is being studied in our latest experiments and appears to be a promising radioprotective agent.)

The toxicity of MPG in combination with AET has not yet been determined exactly, but a relative antagonism between these two radioprotective agents seems to exist.

Pretreatment of mice 10–30 min before irradiation with 0.25 to 2.0 mM/kg (i.e. about 50–350 mg/kg) of MPG has resulted in 30–40% survival of animals irradiated with 630 R of X-rays, or 750 R of cobalt-60 gamma rays, the minimum absolutely lethal doses in our present experimental conditions. Larger amounts of MPG can only provide a lesser degree of survival. No significant protection has been bestowed on mice exposed to supralethal doses of radiation, such as 800 R of X-rays and 900 R of gamma rays or more.

In contrast to the earlier findings with radioprotectors, remarkable therapeutical effects were observed in the survival of irradiated mice when treated with MPG after exposure. A dose of 2 mM/kg of MPG given intraperitoneally 1–5 hours after irradiation with 630 R of X-rays, or

TABLE I. SURVIVAL OF MICE TREATED WITH 2mM/kg (370 mg/kg) MPG INTRA-PERITONEALLY AT DIFFERENT TIMES AFTER IRRADIATION

Time of treatment after irradiation (hours)	630 R of X-rays		700 R of cobalt-60 gamma rays		900 R of cobalt-60 gamma rays	
	n'/n	%	n'/n	%	n'/n	%
1	19/30	43.3	34/48	70.8	16/47	34.0
2	9/30	30.0	34/48	70.8	22/48	45.8
3	17/30	56.7	38/48	79.2	20/48	41.7
4	11/30	36.7	40/48	83.3	28/48	58.3
5	14/30	46.7	27/36	75.0	15/36	41.7
6	–	–	26/36	72.2	9/36	25.0
24	–	–	27/36	75.0	6/36	16.7
Control	0/30	0	14/47	29.8	0/48	0

Note: n' = number of animals surviving 30 days after irradiation;
 n = number of animals irradiated.

TABLE II. SURVIVAL OF MICE TREATED WITH VARIOUS COMBINATIONS OF AET AND MPG 15 min BEFORE EXPOSURE TO COBALT-60 GAMMA RAYS

Dose of AET (mM/kg)	Dose of MPG (mM/kg)	Dose of irradiation (R)							
		900		1000		1100		1200	
		n'/n	%	n'/n	%	n'/n	%	n'/n	%
–	–	0/48	0	0/48	0	0/48	0	0/36	0
0.5	0.25	20/24	83.3	9/12	75.0	0/24	0	0/24	0
0.5	0.5	22/24	91.7	14/22	63.6	6/24	25.0	0/24	0
0.5	1.0	20/24	83.3	16/24	66.7	5/24	20.8	0/24	0
1.0	1.0	24/24	100.0	20/24	83.3	10/24	41.7	3/24	12.5
1.5	2.0	36/36	100.0	36/36	100.0	32/36	88.9	15/24	62.5

Note: 1mM/kg AET = 280 mg/kg; 1 mM/kg MPG = 185 mg/kg.
 n' and n the same as for Table I.

1–24 hours after exposure to 700 and 900 R of gamma rays resulted in a surprisingly high percentage of surviving animals at the 30th day of the experiment (Table I). This favourable effect of MPG has also been confirmed in mice exposed to lethal doses of mixed neutron-gamma radiation generated in a research reactor.

According to this observation, the mechanism of the radioprotective activity of MPG might be quite different from that of other, well-known radioprotective substances.

An enhanced radioprotective effect was obtained when mice, pretreated with a combination of 0.5–1.5 mM/kg of AET and 0.25–2.0 mM/kg of MPG, were exposed to supralethal doses of cobalt-60 gamma radiation in the range of 900–1200 R. The efficiency of radioprotection increased in proportion to the amounts of the constituents in the combination (Table II).

TABLE III. SURVIVAL OF IRRADIATED MICE UNDER THE INFLUENCE OF AET, MPG AND COMBINATIONS OF THEM GIVEN AT VARIOUS TIMES BEFORE OR AFTER IRRADIATION

Time of treatment relative to exposure	Dose of radioprotector (mM/kg)		Survival of mice exposed to:			
			630 R of X-rays		900 R of gamma rays	
	AET	MPG	n'/n	%	n'/n	%
15 min before	–	–	0/15	0	0/48	0
	1.0	–	8/15	53.3	–	–
	–	2.0	5/15	33.3	0/24	0
	0.5	1.0	–	–	20/24	83.3
	1.0	2.0	15/15	100.0	–	–
	1.5	2.0	15/15	100.0	36/36	100.0
60 min after	1.0	–	0/20	0	–	–
	–	1.0	–	–	9/24	37.5
	–	2.0	6/15	40.0	11/23	47.8
	–	3.0	7/15	46.7	–	–
	1.5	2.0	–	–	0/12	0
15 min before and 60 min after	1.0	–	10/15	66.7	–	–
	–	2.0				

See footnotes to Tables I and II.

As expected, no radioprotective effect was observed when AET was given in combination with MPG after irradiation. However, the same combination proved to be effective when AET was given before irradiation, as usual, and MPG was given after irradiation. However, the efficiency of this pre- and post-treatment was not significantly higher than that of AET given alone before exposure or that of MPG given alone after irradiation (Table III).

On the basis of these rather preliminary results, our impression is that the radioprotective effect of AET and MPG given in combination before irradiation is much higher than would be expected from the added effects of the two radioprotective substances of different mechanisms of action. However, the potentiating character of synergism must be verified unambiguously in the course of further quantitative analyses.

REFERENCES

[1] DOHERTY, D.G., BURNETT, W.T., Jr., Protective effect of S, β-aminoethylisothiuronium Br. HBr and related compounds against X-radiation death in mice, Proc. Soc. Exp. Biol. **89** (1955) 312.
[2] DOHERTY, D.G., BURNETT, W.T., Jr., SHAPIRA, R., Chemical protection against ionizing radiation. II. Mercaptoalkylamines and related compounds with protective activity, Radiat. Res. **7** (1957) 13.
[3] SHAPIRA, R., DOHERTY, D.G., BURNETT, W.T., Jr., Chemical protection against ionizing radiation. III. Mercaptoalkylguanidines and related isothiuronium compounds with protective activity, Radiat. Res. **7** (1957) 22.
[4] SZTANYIK, L., " Studies on chemical radiation protection" (Proc. Symp. on Radiation Damage – Radiation Protection), (FRIED, L., UJHELYI, A., Eds), ORSI, Budapest (1959) 214 (in Hungarian).
[5] DOHERTY, D.G., "Chemical protection to mammals against ionizing radiation", Radiation Protection and Recovery (HOLLAENDER, A., Ed.), Pergamon Press, Oxford (1960) 45.

[6] SZTANYIK, L., "Chemical radioprotection", Ch. 13, Radiobiology (VÁRTERÉSZ, V., Ed.), Medicina, Budapest (1963) 399 (in Hungarian).
[7] SZTANYIK, B.L., Data on Radioprotective Effect of AET and Structurally Related Compounds, Thesis, Budapest (1965) 1–376 (in Hungarian).
[8] SZTANYIK, L., VÁRTERÉSZ, V., DÖKLEN, A., NÁDOR, K., Studies on the comparison of the radioprotective effect of AET and its cyclic analogues, Prog. Biochem. Pharm. 1 (1965) 515.
[9] SZTANYIK, L., "Chemischer Strahlenschutz", Strahlenbiologie (VÁRTERÉSZ, V., Ed.), Akadémiai Kiadó, Budapest (1966) 495.
[10] SZTANYIK, L., "Studies on the correlation between the chemical structure and radioprotective effect of AET-type compounds", 10th Anniversary of the National Research Institute for Radiobiology and Radiohygiene (VÁRTERÉSZ, V., Ed.), Medicina, Budapest (1967) 56 (in Hungarian).
[11] RUSANOV, A.M., ALEXEYEVA, G.N., KOLESOVA, M.B., SZTANYIK, L., NÁDOR, K., GYÖRGY, L., Radioprotective properties of alkyl-, aminoalkyl- and cyclic-derivatives of AET, Vopr. Radiobiol. Roentgeno-Radiol. 6 (1968) 124 (in Russian).
[12] VÁRTERÉSZ, V., SZTANYIK, L., NÁDOR, K., "Radioprotective effect of N-substituted AET-derivatives having amino acid structure", Radiation Protection and Sensitization (MOROSON, H.L., QUINTILIANI, M., Eds) Taylor and Francis, Ltd., London (1970) 325.
[13] SÁNTHA, A., VÁRTERÉSZ, V., MÁNDI, E., NÁDOR, K., ZARÁND, P., Radioprotective effect of new xanthogenic acid derivatives on mice exposed to mixed reactor radiation, Adv. Antimicrobial and Antineo-plastic Chemotherapy, Avicenum, Prague (1972) 843.
[14] FASTIER, F.N., Structure-activity relationships of amidine derivatives, Pharmacol. Rev. 14 (1962) 37.
[15] VAN BEKKUM, D.W., The protective action of dithiocarbamates against the lethal effects of X-irradiation in mice, Acta Physiol. Pharmacol. Neerl. 4 (1956) 508.
[16] DiSTEFANO, V., LEARY, D.E., DOHERTY, D.G., The pharmacology of β-aminoethylisothiuronium bromide in the cat, J. Pharmacol. Exp. Ther. 117 (1956) 425.
[17] KOCH, R., SCHWARZE, W., Toxikologische und chemische Untersuchungen an β-aminoäthylisothiuronium-Verbindungen, Arzneim.–Forsch. 7 (1957) 576.
[18] SZTANYIK, L., GYÖRGY, L., Investigations with radioprotective substances. II. Pharmacological effect of AET, Honvédorvos 11 (1959) 285 (in Hungarian).
[19] MARMO, E., IMPERATORE, A., DI GIACOMO, S., Effetti della metoclopramide sull'apparato cardiovascolare e sul relativo SNV, Clin. Terap. 51 (1969) 509.
[20] SÁNTHA, A., SZTANYIK, L., "Diminution of toxic side effects of AET with pharmacological antagonists", Experimental Studies on the Early Radiation Reactions 2 (1970) 172 (in Russian).
[21] SÁNTHA, A., SZTANYIK, L., "Study on the cardiotoxic effects of radioprotective compounds and their pharmacological prevention in irradiated animals", Proc. 1st Congr. Hung. Pharmacol., Akadémiai Kiadó, Budapest (1972) 257
[22] SZTANYIK, B.L., MÁNDI, E. (unpublished data).
[23] WANG, R.I.H., KEREIAKES, J.G., Increased survival from radiation by mixtures of radioprotective compounds, Acta Radiol. 58 (1962) 99.
[24] GANTZ, J.A., WANG, R.I.H., Reduction in radiation lethality by chemical mixture and bone marrow in mice, J. Nucl. Med. 5 (1964) 606.
[25] MAISIN, J.R., MATTELIN, G., Reduction in radiation lethality by mixtures of chemical protectors, Nature, London 214 (1967) 207.
[26] SZTANYIK, B.L., VÁRTERÉSZ, V., "Radioprotective effect of a mixture of AET and 5-methoxytryptamine in X-irradiated mice", Radiation Protection and Sensitization (MOROSON, H.L., QUINTILIANI, M., Eds), Taylor and Francis, Ltd., London (1970) 363.
[27] SZTANYIK, B.L., SÁNTHA, A., VÁRTERÉSZ, V., Quantitative investigations into the synergistic effect of radioprotective substances, 4th Int. Congr. Radiat. Res., Evian (1970), Abstr. No. 834.
[28] SÁNTHA, A., SZTANYIK, L., MÁNDI, E., Investigations into the effects of combined radioprotective substances in animal experiments, Honvédorvos 24 (1972) 279 (in Hungarian).
[29] GRAY, J.L., TEW, J.T., JENSEN, H., Protective effect of serotonin and para-aminopropiophenone against lethal doses of X-irradiation, Proc. Soc. Exp. Biol. 80 (1952) 604.
[30] DUKOR, P., Versuche zum Mechanismus der Strahlenschutzwirkung von Oxytryptaminderivaten, Strahlentherapie 117 (1972) 330.
[31] KRASNIH, I.G., ZHEREBCHENKO, P.G., Radioprotective effect of 5-methoxytryptamine and other alkoxy derivatives of tryptamine, Radiobiol. 2 (1962) 156 (in Russian).
[32] FEHÉR, I., GIDÁLI, J., SZTANYIK, L., Study of the radioprotective effect of 5-methoxytryptamine on haemopoietic stem cells, Int. J. Radiat. Biol. 14 (1968) 257.
[33] PATT, H.M., TYREE, E.B., STRAUBE, R.S., Cysteine protection against X-irradiation, Science 110 (1949) 213.

[34] SUGAHARA, T., TANAKA, Y., NAGATA, H., TANAKA, T., KANO, E., "Radiation protection by 2-mercaptopropionylglycine" (Proc. Int. Symp. on Thiola), Santen Pharmaceutical Co., Ltd. (1970) 267.
[35] SÁNTHA, A., MÁNDI, E., BENKÓ, G., Studies on mercaptopropionylglycine, a new radioprotector, Honvédorvos **26** (1974) 145 (in Hungarian).

INTERFERENCE WITH ENDOGENOUS RADIOPROTECTORS AS A METHOD OF RADIOSENSITIZATION

S. ŁUKIEWICZ
Institute of Molecular Biology,
Jagiellonian University,
Cracow, Poland

Abstract

INTERFERENCE WITH ENDOGENOUS RADIOPROTECTORS AS A METHOD OF RADIOSENSITIZATION.
The concept of a radioprotective function of natural melanins is discussed in the light of some earlier studies and new experimental findings. Data are presented which indicate the possibility of producing a marked radiosensitization in a variety of biological systems, including animal and human melanoma, by an appropriate interference with the melanin system. The perspectives of future radiotherapeutical applications are pointed out, taking as a basis the positive results of exploratory clinical trials.

1. INTRODUCTION

An inherent level of radiosensitivity in biological systems can be modified in several ways. One of the possible approaches is based on the idea of either enhancing or counteracting the normal functions of the endogenous radioprotectors, i.e. those natural cell constituents which are essential for cellular radioresistance.

One can a priori expect the radioprotection or radiosensitization brought about by applying this principle to be the more prominent the higher the concentration of natural radioprotectors present in a given type of cells. This holds true, for example, for endogenous thiol compounds and melanins which sometimes accumulate in the cytoplasm in quite substantial amounts. In such cases the natural radioprotectors have a chance of predominating among the factors which determine the overall level of radiosensitivity. *Micrococcus radiodurans* is probably a good example of this kind of relation, since its extreme radioresistance has been shown to depend mainly on a high content of endogenous thiols [1]. If, however, the SH-groups of these cell components are blocked with p-hydroxymercuribenzoate a drastic radiosensitization can be observed [2].

The melanin-containing cells are believed to owe their considerable radioresistance to the radioprotective properties of the pigment [3]. Thus, the question may arise whether the normal radiation response of pigmented cells of plant or animal origin can be modified by affecting the amount or physico-chemical features of melanin.

Two possibilities can be taken into account in this regard: the first, that radioresistance may be enhanced by favouring the radioprotective activity of the pigment; the second that radioresistance may decrease if the protective capacity of this polymer can be in some way weakened or eliminated

The present paper aims at:

(a) Giving a brief account of an earlier work pertaining to these problems;
(b) Discussing some further experimental evidence in favour of the concept that melanins may act as radioprotectors in both model and living systems;

(c) Demonstrating that the interference with melanogenesis may be followed by a significant radiosensitization of plant and animal cells which is strong enough to open the way to useful radiotherapeutical applications.

2. RADIOPROTECTIVE ROLE OF MELANIN

An exceptionally high accumulation of paramagnetic centres in the molecules of melanin is considered by some authors to be a good reason for assuming that this polymer should be able to interact with radiation-induced free radical species in a manner resembling common radical scavengers [4]. On the other hand, quite considerable stability of the paramagnetic properties of melanin is also well known: they can persist even after such a drastic treatment as, for example, boiling melanin with concentrated hydrochloric acid. This and other data might be taken to mean that a substantial number of the melanin centres are probably hidden deep inside the structure of the polymer, thus being poorly accessible to free radicals interacting with the "surface" of a giant macromolecule of the pigment [5].

The above reasons lead some authors to believe [6] that at the present stage it appears premature to accept as a proven fact the view ascribing the radioprotective properties of melanins to their strong paramagnetism. On the contrary, data are available which clearly indicate that melanin may sometimes reduce the amplitude of EPR signals of certain spin labels without any measurable decrease in its own content of unpaired electrons [7].

Thus, the process of recombination need not necessarily be involved in the interactions of free radical compounds with the melanin centres and the latter may prove to be of secondary importance for the activity of the polymer as a radical scavenger. More work, therefore, is needed before one can provide a better interpretation of the molecular mechanisms underlying the radioprotective properties of melanin.

At the same time there is an increasing bulk of evidence based on both theoretical considerations and empirical findings that melanins are able to influence the level of cellular radiosensitivity. Some examples of such data are discussed in Section 2.1. In general, the current situation might perhaps be briefly characterized by stating that:

— classification of melanins as natural radioprotectors seems to be justified for a good many reasons,
— the exact way in which this peculiar property is realized by melanins is not, however, yet understood in full detail.

2.1. Selected data from earlier studies

The concept of radioprotection exerted by melanins is based on three groups of premises:

(a) The assumption originating from theoretical, quantum mechanical studies on the electronic structure of melanins that this biopolymer has the capacity of an exceptionally strong electron acceptor [8–10];
(b) The conclusions from physical experiments on simple, non-living model systems, pointing to a possible regulatory action of melanins, postulated to be a sort of control mechanism maintaining the concentration of unpaired electrons and of the oxidation-reduction potential of cytoplasm at a balanced level [11–13];
(c) The results of comparative radiobiological investigations indicating the existence of a relationship between the concentration of pigment in living objects and their radiation responses [14, 15].

Only a limited number of publications pertaining to the problems listed under items (b) and (c) are briefly discussed in this section, in view of the fact that several comprehensive monographs and reviews are available which deal with many of the physical and biological aspects of melanin and pigment cells [16–21], including radioprotection.

Two papers describing model experiments appear to be especially significant in the context of the present discussion. The first is the study by Dain et al. [13] on the interactions of melanin with free radicals induced by u.v. illumination in frozen solutions of proteins. The second is the analysis of free radical generation in pigment granules by visible light [22]. It is important that the scope of the above investigations has not been limited to the mere generation of unpaired electrons by light or u.v. illumination but has been extended to the phenomena of "uptake" by this polymer of similar photo-induced electrons from its nearest molecular surroundings. This kind of interaction was thought to prevent some of the harmful effects of u.v. illumination and light.

If melanin is indeed capable of such "trapping" then one can reasonably expect the existence of a similar mechanism in the case of free radical products of radiolytic origin. A number of observations on model systems and living objects have been reported [3, 14, 15, 23] which seem to confirm this supposition (see also Section 2.2.1).

The striking radioresistance of *Nadsoniella nigra* has been ascribed to large deposits of melanin found in the cells of this antarctic species of fungus [24]. To check whether this inference is correct, the radiosensitivity of several species belonging to the family *Dematiaceae* was compared and correlated with the melanin content. It could be revealed in this way that the higher the accumulation of pigment (as determined by means of EPR spectroscopy) the greater the radioresistance of a given species [15]. A weak point of such an approach, however, results from the fact that various species, whether pigmented or not, may differ in their radioresistance, even if they are phylogenetically related. Thus, it is not easy to establish the extent to which pigmentation contributed to the observed differences in radiation response.

To avoid this difficulty, biological materials with as similar as possible genetical and histological features should be chosen for comparison. An example of this kind of study is the work of Prasad and Johnson [14] who examined the hepatic melanocytes of *Amphiuma tridactylum* known to contain a varying amount of melanin. The authors were able to show that the lightly pigmented cells were more sensitive to X-rays than those with abundant pigment.

Distinct differences in melanin concentration could also be noted occasionally among the cells of S 91 Cloudman melanoma in mice. Cobb [25] demonstrated in 1956 that pale slices of neoplastic tissue irradiated in vitro were more susceptible to X-ray damage than explants with an intense pigmentation.

The results of radiotherapy in 135 cases of human melanoma lead to a similar conclusion: weaker response to radiation used to be associated with the high degree of melanization [26]. Unfortunately, an exact estimation of radiosensitivity of tumour tissue in individual patients is a rather complicated problem and, moreover, radioresistance is only one of the many factors which determine the final effects of treatment. Therefore, poor results in radiotherapy need not necessarily mean a high radioresistance of the tumour. Nevertheless, the opinion that the low curability of human melanoma is mainly due to its low sensitivity to ionizing radiations is much more common. An opposite standpoint was taken by Dewey [27] who, however, does not seem to have many supporters. Finally, it should be pointed out that more recent studies on transplantable hamster melanoma carried out on tumours growing in situ or cultured in vitro (see Section 2.2.2) are in agreement with the above-mentioned clinical observations [26], and not with Dewey's claims.

To sum up, it may be said that in many instances, whether dealing with living or non-living systems, with plant, animal, or human tissues, normal or pathologically altered, one often comes across situations indicating that the amount of melanin present in the system appears to influence its reactivity to radiation.

A quite consonant inference follows from the experimental work described in Section 2.2.

2.2. New experimental data

This section is devoted to a discussion of some recent investigations which are a continuation of the two lines of research, as distinguished in the items 2.1(b) and 2.1(c).

Although these studies were performed on quite different subjects, i.e. either non-living (2.2.1) or living (2.2.2) systems, they have a common leading idea: they were undertaken in an attempt to analyse the radioprotective role of melanins, and in particular the significance of the amount of this polymer contained in the system at or shortly after radiation exposure.

2.2.1. EPR studies on model systems

The work of Dain et al. [13] referred to in Section 2.1 provided very instructive results pertaining to the behaviour of a solid-state model (frozen aqueous solutions of protein mixed with melanin and examined at liquid nitrogen temperature) under u.v. illumination. This kind of analysis has been extended and modified by applying X-rays instead of u.v., and by using another solid-state system suitable for EPR study at room temperature [23].

A very convenient model proved to be a dry melanin or melano-proteid fixed in polymethylmetacrylate, the latter being an organic solid-state matrix. The advantage of choosing this particular polymer as a matrix lies in the circumstance that the X-ray-induced modifications in its molecular structure can be easily detected by means of EPR spectroscopy without the necessity of freezing the system down to liquid nitrogen temperature.

X-ray exposures to doses in the range of a few thousand rads result in the appearance of the asymmetric single EPR line shown in Fig.1. It has a width of about 4 G and a value of the spectroscopic splitting factor "g" near 2.0. This signal is due to the formation of radicals of the type (I) with an unpaired electron located at the carboxylic group [28, 29].

$$\sim CH_2 - \underset{\underset{COO^{\cdot}}{|}}{\overset{\overset{CH_3\ (H)}{|}}{C}} \sim \qquad\qquad \sim CH_2 - \underset{\underset{\underset{O\quad OCH_3}{\diagup\ \diagdown}}{C-CH_3}}{\overset{\overset{CH_3\ (H)}{|}}{C}} \sim$$

$$(I) \qquad\qquad\qquad\qquad (II)$$

Upon exposures long enough to exceed a dose of 7500 rad another EPR signal gradually becomes visible. Its shape can be seen in Fig.2. It is composed of the seven lines of hyperfine structure with intensities 2 : 8 : 1 : 12 : 1 : 8 : 2, and a width of 7.5 G. A radical represented by formula (II) is responsible for this EPR line [28–30].

Hence, it is evident that EPR spectroscopy is a very suitable tool for this kind of study as it offers an excellent insight into the molecular rearrangements within the irradiated sample. The formation of these radicals results from the radiolytic breakdown of certain bonds. Therefore, the accumulation of paramagnetic products in the polymer reflects the degree of "damage" to its structure brought about by irradiation. On the other hand, the actual concentration of radicals can be easily monitored as it changes during X-ray exposure, simply by looking at the amplitude or intensity of an X-ray-induced signal regularly increasing up to a certain level. Figure 3 illustrates the kinetics of this process.

The radicals described, however, are not stable at room temperature. Their number starts to decrease as soon as radiation exposure is interrupted. The decay due to recombination in a solid-state matrix is not a speedy process, as can be seen from Fig.4. It may last for many hours or even days, and can be followed by watching continuously the EPR spectrum of the irradiated

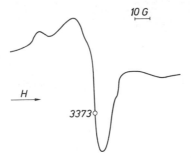

FIG.1. *An asymmetrical EPR line observed after X-raying a sample of pure polymethylmetacrylate with a dose lower than 7500 rad.*

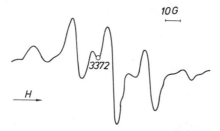

FIG.2. *An EPR signal induced in a sample of pure polymethylmetacrylate by X-irradiation with doses exceeding 7500 rad. Note the seven lines of hyperfine structure.*

sample. In general, the paramagnetic species responsible for the signal shown in Fig.2 are much more stable than those giving rise to a single EPR line. Thus, the first of them are able to persist for periods as long as one or two weeks in massively irradiated samples whereas the second ones may become undetectable within 24 hours after the end of exposure [23].

The above relations drastically change upon the addition of melanin, even if these are very small amounts — less than 2% of the total mass of the matrix. First of all, the radiative formation of radicals occurs with a much lower yield. In other words, concentrations of X-ray-induced unpaired electrons, corresponding to a given dose of radiation, are smaller in a model "melanin-matrix" than in a pure matrix. This results in a less intense production and accumulation of radicals, as reflected by the flattening of curves which depict the increase in the EPR absorption during irradiation (see Fig.3). Consequently, the dose required to reach a certain level of EPR absorption is greater in models dotted with melanin. The effect clearly depends upon the amount of melanin added, as is evident from the diagram in Fig.5.

The decay time of X-ray-induced EPR signals is also strongly affected by the presence of melanin in the system: the pigment shortens the period necessary for the radicals to recombine and for the EPR absorption to disappear. This is visible in Fig.4 whereas Fig.6 shows the effect to be influenced by the concentration of pigment introduced into the matrix. The reduction of the decay time may be very striking, sometimes three-fold.

The change in the decay time of the X-ray-induced EPR signals is often quite drastic. One may have the impression that unpaired electrons were "sucked in" by the melanin in a way resembling the soaking of ink into blotting paper. Such a presentation must lead to a natural question as to the presumable fate of these electrons. Do they recombine with the melanin centres or are they trapped in some way by the biopolymer? In the first case a decrease in the amplitude of EPR signal of melanin should be observed. The validity of the second supposition would mean that an opposite change, i.e. an increase in the signal intensity of melanin, should be

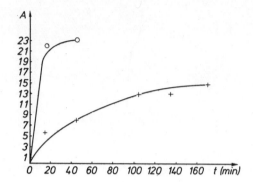

FIG. 3. The increase in the amplitude of an X-ray-induced EPR signal (A) as a function of exposure time. The values of "A" are given in relative units; time in minutes.
○ — pure polymethylmetacrylate
x — model system: organic matrix with melanin added.
Mean dose-rate about 3000 rad/min.

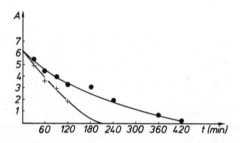

FIG. 4. The decrease in the amplitude of an X-ray-induced EPR signal (A) as a function of decay time. The values of "A" are given in relative units; time in minutes.
○ — pure polymethylmetacrylate
x — model system: organic matrix with melanin added.
All observations were made at room temperature.

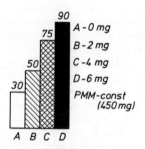

FIG. 5. Diagram illustrating the changes in the level of X-ray doses necessary to induce an EPR signal of the same standard amplitude in model systems with varying amounts of melanin.
A — 450 mg of pure polymethylmetacrylate
B — 450 mg of pure polymethylmetacrylate with 2 mg of melanin
C — 450 mg of pure polymethylmetacrylate with 4 mg of melanin
D — 450 mg of pure polymethylmetacrylate with 6 mg of melanin.

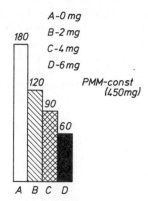

FIG. 6. *Diagram illustrating the decay time of an EPR signal induced by X-irradiation of a model system composed of polymethylmetacrylate with varying amounts of melanin added. The letters mean the same as in Fig. 5.*

expected. Instead, neither of these events takes place in the experiment: the X-ray-induced radicals are "neutralized" by melanin without any measurable modification of its own EPR signal. This might indicate that the interactions of pigment with the radiolytic products are mainly due to oxidation-reduction processes in which the "natural" pool of the melanin paramagnetic centres does not seem to be engaged.

The lower yield of X-ray-induced radicals and their more rapid decay may be interpreted as the manifestations of a "protective action" effected by melanin addition to the matrix.

2.2.2. Comparative study of radiosensitivity in hamster melanoma

The transplantable hamster melanoma has two stable lines: a melanotic and amelanotic one. The first is characterized by an abundant pigmentation, while the cells of the second do not contain any detectable amount of melanin.

This spontaneous defect in melanin biosynthesis provides a good opportunity for testing whether the difference in melanin content is associated — in this particular case too — with unequal radiation responses. One can a priori expect on theoretical grounds that the amelanotic form of tumour should be more sensitive than its melanotic counterpart.

In fact, the careful comparison of reactivity to X-rays which was made for both in-vitro and in-situ conditions of cell growth confirmed this prediction.

2.2.2.1. In-vitro experiments with Bomirski melanoma

In the exploratory series of determinations, stabilized lines of cells were, unfortunately, not available for the in-vitro culture. Hence, quite heterogeneous cell populations obtained by trypsin digestion of tumour explants had to be used. As a criterion of radiation damage, cell survival after a period of 10 days, evaluated by means of vital staining, was applied.

In this way it could be revealed that the melanotic form is nearly twice as resistant as the amelanotic one, LD_{50} amounting to 750 rad and 440 rad, respectively [31]. The described methodical approach is, however, far from reliable. Nevertheless, additional work recently done with more uniform cell populations, stabilized in the conditions of tissue culture, produced an entirely consonant result. The capacity of forming clones was applied in this case as a criterion for estimation of lethal doses [32].

It might be interesting to note that a longer maintenance of melanotic cells in vitro brings about, as a rule, their gradual and almost complete depigmentation. Thus, for example, the strains of human melanotic melanoma M22 and M81 do not contain after 184 passages even the smallest traces of pigment detectable in their EPR spectra. Their radioresistance, as measured at this stage, is close to the level of LD_{50} characteristic of amelanotic hamster melanoma, being only a half of the value established for the melanotic form. In other words, these two lines became quite sensitive to X-rays after a long in-vitro cultivation. This effect is associated with a complete loss of pigment [33].

Finally, it proved possible to separate viable hamster melanoma cells with varying amounts of melanin by centrifuging their suspension in a density gradient, and to compare the radiosensitivity of various fractions. The melanoma cells with a medium melanin concentration revealed an intermediate value of LD_{50}, higher than that of amelanotic cells but lower than the LD_{50} of cells with an intense pigmentation [32].

Hence, the data presented in this paragraph appear to be some further evidence in favour of a causal relationship between pigmentation and radiation responses.

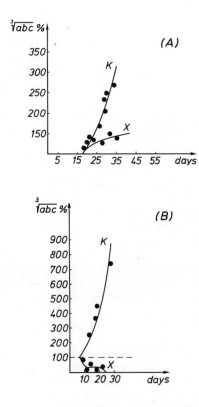

FIG. 7. The growth and regression curves of (A) a melanotic form and (B) an amelanotic form of Bomirski melanoma in the Syrian hamster with and without irradiation.
$\sqrt[3]{abc}$ *taken as a measure of tumour size (for explanation see text).*

  K — non-irradiated tumours
  X — irradiated tumours

2.2.2.2. In-situ growing of Bomirski melanoma

Very remarkable differences in radiosensitivity between melanotic and amelanotic forms could also be demonstrated in tumours implanted into the skin of the Syrian hamster [31]. X-irradiation (50 kV) with a total dose of about 5000 rad, fractionated over a period of one month, proved to be efficient enough to cause a complete regression of amelanotic tumours within 3—4 weeks. Non-irradiated tumours increase their volume about ten-fold during the same period (see Fig. 7B, showing the curves of growth of an amelanotic form of Bomirski melanoma). The size of tumours was estimated according to Schreck's method by measuring their three main dimensions and calculating the value of $\sqrt[3]{a \cdot b \cdot c}$ where a, b and c are the dimensions of the tumour along three axes (length, width and thickness).

There is no doubt in the light of these observations that the amelanotic form is fairly radiosensitive in contrast to its pigmented counterpart. Figure 7A illustrates the rate of growth of irradiated and non-irradiated melanotic tumours.

One can see from Fig. 7A that a distinct growth retardation can be brought about in the pigmented Bomirski melanoma (by a factor of approximately two). Total regression or irreversible inhibition of growth could, however, never be observed for melanotic tumours upon application of the same doses which always produced this effect in the amelanotic form of hamster melanoma. These differences are clearly visible in Fig. 8. The rates of growth or regression are represented by the slope of the lines (or values of the regression coefficients). Thus, the angle between these lines reflects the changes due to irradiation. It is evident that the radiation response of the amelanotic (X_{AM}) form is several times stronger than that of the pigmented tumours (X_M). The denotations K_{AM} and K_M pertain to amelanotic and melanotic tumours, respectively, growing in the control animals.

The comparative studies of radiosensitivity described in 2.2.2.1 and 2.2.2.2 gave consistent results indicating that the amelanotic form of Bomirski hamster melanoma is considerably more sensitive to X-irradiation than the melanin-containing tumour, both in vitro and in situ.

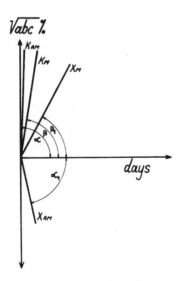

FIG. 8. Diagram illustrating the difference in radiosensitivity of melanotic and amelanotic Bomirski melanoma in hamsters (for explanation see text).

This conclusion is in agreement with the general picture emerging on the basis of data described in Sections 2.1. and 2.2.1. One can observe that the decrease in melanin content of a system (living or non-living) is usually followed by an enhancement of radiation responses.

The cases of depigmentation mentioned in 2.2.2.1 and 2.2.2.2 occur "spontaneously", either as a result of the action of a natural tyrosinase inhibitor [34], which is probably responsible for the appearance of amelanotic forms of melanoma, or under the influence of certain not precisely identified environmental factors in the in-vitro culture [35]. The question may, however, be put as to whether experimentally provoked diminution in melanin accumulation in the cytoplasm will also bring about similar radiobiological consequences in the form of a sensitization to radiation.

The following sections contain a discussion of this problem.

3. RADIOSENSITIZATION BY INTERFERENCE WITH THE MELANIN SYSTEM

The data discussed in Section 2 aimed at demonstrating that the concept of radioprotection exerted by melanins has received substantial support from the experiment. Thus, assuming that this biopolymer is indeed able to act as a natural radioprotector, one can suggest at least three different ways of reducing its protective capacity:

(a) Modification of those physico-chemical features of melanin molecules which are responsible for their activity as a protector;
(b) Inhibition of the biosynthesis of the pigment in an attempt to decrease its amount in the cell;
(c) Simultaneous action affecting both the properties and concentration of melanin.

3.1. Modification of the melanin properties

As pointed out in Section 2, there is no agreement as to the question which of the physical or chemical features endows melanins with their capacity of protection against ionizing radiation.

Some authors believe that the free radical property is most important from this point of view [4, 12]. On the basis of this assumption, attempts were made to modify the radiosensitivity of hamster melanoma by the pretreatment of tumour-bearing animals with chlorpromazine [4]. This compound is an excellent electron donor and is known to form charge transfer complexes with melanins [36]. These complexes are associated with the change in paramagnetic properties of the pigment. The authors [4] noted that chlorpromazine is able to produce a radiosensitization which leads to decreased transplantability and a lower incidence of metastases as well as to a retardation of growth (the size of the tumours being 61% smaller in X-irradiated and pretreated animals). Nevertheless, more extensive work is still needed since the number of experimental animals used in the study was, unfortunately, rather limited [15] and no parallel observations with the amelanotic form of melanoma were carried out. In any case, a possible future confirmation of these results by additional experiments will not necessarily mean that the effect was due to the change in free radical properties of the pigment. Another interpretation of such a sensitization also seems plausible. Chlorpromazine is able to bind metal ions such as Cu^{2+}, which is indispensable for the normal function of tyrosinase. For this reason, the observed phenomena can also be caused by the interference with melanin biosynthesis (see Section 3.2) or by influencing both the amount and properties of pigment.

3.2. Interference with melanin biosynthesis

The spontaneous block of tyrosinase activity, involving the participation of some natural inhibitor and resulting in the formation of amelanotic forms among plants and animals, is a fairly rare phenomenon which cannot be controlled at will by an experimenter.

Therefore, it appeared worth while to check the possibility of imitating these natural "defects" by applying some exogenous tyrosinase inhibitors, and to see whether any change in radiosensitivity would result from such treatment. It should be emphasized, however, that this approach was primarily dictated by the intention of modifying the **amount** of melanin produced by the cell. The main purpose of the adopted procedure was, in other words, an attempt to obtain experimentally "amelanotic" (pigment-deficient) forms of plant and animal cells similar to those occurring occasionally in nature as spontaneous mutants.

The data presented in the following sub-sections are in agreement with the above-formulated expectations. They indicated, moreover, that the mere presence or absence of melanin in the cytoplasm is not sufficient in itself to determine the level of radiosensitivity of pigmented cells. The chemically induced sensitization may take place not only in cells totally deprived of pigment but also in those with a partial depigmentation. This means that apart from the amount and properties of melanin, an undisturbed activity of the whole pigment-producing system of the cell should also be taken into account.

3.2.1. Chemically induced radiosensitization of pigmented plant cells

The growing mycelia of *Aspergillus ornatus*, which gradually develop their pigmentation within 7–10 days after inoculation of spores (see Fig.9), proved to be a convenient material for studying radiation responses. The biological effect of X-irradiation can be easily evaluated by measuring the rate of growth of mycelia.

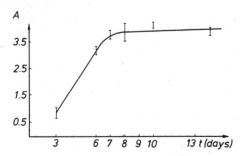

FIG.9. Changes in the amplitude of EPR signals (A) of the growing Aspergillus ornatus mycelia reflecting the gradual increase in their melanin content. Values of "A" given in relative units calculated per gram of dry mass of mycelium. Time of culturing (without addition of any inhibitors) given in days.

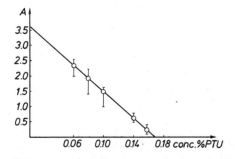

FIG.10. Relationship between the amplitude of EPR signals (A) of Aspergillus ornatus mycelium at the 7th day of culture and the content of phenylthiourea (% PTU) in the medium. The amplitudes of the EPR signals reflect the degree of depigmentation of the mycelia as a result of inhibited melanogenesis. Values of "A" are given in relative units calculated per gram of dry mass of mycelium.

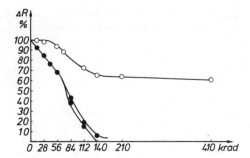

FIG.11. *The growth inhibition of* Aspergillus ornatus *mycelia produced by various doses of X-rays in the presence or absence of tyrosinase inhibitor in the medium. Growth measured as an increase per cent in the diameter of mycelia.*
○ — *control cultures without addition of an inhibitor*
● — *experimental cultures containing 0.16 or 0.18% phenylthiourea in the medium.*

The normal production of pigment was inhibited by means of phenylthiourea. In this way it was possible to bring about a spectacular depigmentation of cells. The loss of melanin could be monitored by recording the amplitude of the melanin EPR signal. The degree of depigmentation was found to depend on the concentration of the inhibitor in the medium (see Fig.10). The mycelia cultivated in the presence of 0.18 or 0.16% phenylthiourea (PTU) did not show any detectable amount of melanin on the 7th day of their growth. Comparison of the curves of growth of normal and pigment-deficient mycelia after irradiation with various doses of X-rays showed that the increasing growth inhibition was correlated with a diminishing melanin content of cells. At higher dose levels the radiosensitivity was seen to be enhanced more than three-fold in the cells with an inhibited melanogenesis. Thus, a total growth inhibition could be obtained in phenylthiourea-treated colonies by irradiation with about 130 000 rad, whereas the control mycelia were able to grow at a slightly lower rate even after receiving doses as high as 430 000 rad (see Fig.11).

A certain disadvantage in using phenylthiourea as a tyrosinase inhibitor lies in its moderate toxicity. Therefore, a systematic study was undertaken [37] to search for other substances which might affect melanogenesis without measurable changes in the growth of mycelia. Substances more effective as inhibitors and at the same time less toxic than phenylthiourea were found. Some of them proved to be good radiosensitizers for *Aspergillus ornatus*. More details pertaining to this work will be published elsewhere [37].

3.2.2. Radiosensitization of amphibian embryos

Another series of experiments using the same tyrosinase inhibitor was carried out on the early embryos of *Xenopus laevis* whose cells are also known to contain abundant melanin pigmentation [38]. Advantage was taken of the method developed by Dumont and Eppig [39] which makes it possible to produce unpigmented *Xenopus* eggs capable of normal development.

The melanogenesis was blocked during the period of egg maturation in female frogs by daily injections of 0.4% phenylthiourea for one month or more. The eggs from successive ovulations (stimulated by the administration of gonadotropins) contained a gradually diminishing amount of pigment. The pigment could be determined quantitatively by measuring the EPR spectra of the eggs (see Fig.12). The curve depicting the kinetics of depigmentation had a somewhat different shape for developing embryos, as shown in Fig.13, but the general tendency was the same so that after a few days of treatment a marked decrease in the EPR signal amplitude was always observable.

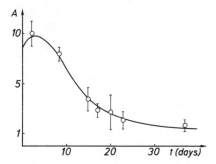

FIG.12. The EPR signal amplitude of Xenopus laevis eggs as measured at the successive ovulations during a period of 40-days treatment of female frogs with phenylthiourea. The value of amplitude "A", given in arbitrary units, reflects the melanin content of egg cells which is gradually diminishing as a result of tyrosinase inhibition.

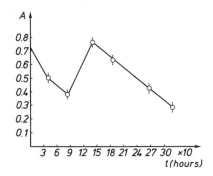

FIG.13. The changes in the EPR signal amplitude of Xenopus laevis embryos observed during the treatment of female frogs with phenylthiourea. Time of tyrosinase inhibition given in hours. The values of "A" are expressed in relative units.

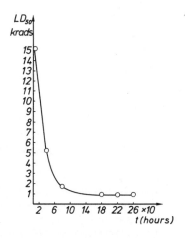

FIG.14. The changes in radiation response of Xenopus laevis embryos during the treatment of female frogs with phenylthiourea. Radiosensitivity expressed as LD_{50} in kilorads. Time of tyrosinase inhibition given in hours.

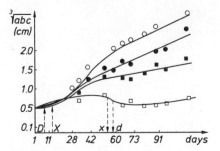

FIG.15. The curves of growth of Bomirski melanotic melanoma implanted into the skin of Syrian hamsters. The tumour sizes are expressed as $\sqrt[3]{a\,b\,c}$ (for explanation see text).
○ — *control experiment (no drug, no irradiation)*
● — *drug treatment without irradiation*
■ — *irradiation without drug treatment*
□ — *irradiation with a parallel drug treatment.*
The moments of starting or ending drug administration and X-irradiation are marked by arrows.

Irreversible inhibition of development (at stage No. 12 according to the classification of Nieuwkop and Faber) or the lethal effect after 48—72 hours were adopted as criteria of radiation damage. Radiosensitization was found to develop before the complete loss of melanin (see Fig.14). Thus, full depigmentation was not an essential condition for the appearance of reduced radioresistance (cf. Figs 13 and 14). Sensitization could be noted not only among pigment-deficient embryos (at later stages of melanogenesis inhibition) but also among those which contained abundant amounts of melanin, i.e. after a few days of treatment [40]. This might mean that the pigment can lose its protective capacity in certain circumstances or that a normal activity of the melanin-producing enzymatic system is a prerequisite of radioprotection. This problem, however, requires further study.

3.2.3. Radiosensitization of Bomirski hamster melanoma

A number of substances, including phenylthiourea, were tested for the capacity of sensitizing hamster melanome against radiation [41, 42]. All the examined compounds were believed to affect the level of pigmentation, though no marked changes in melanin content were macroscopically visible. In any case, a total depigmentation comparable with the phenomena described in Sections 3.2.1 and 3.2.2 (see also Refs [37] and [40]) was never achieved: tumours always remained intensely pigmented. Nevertheless, the radiation responses of tumours growing in the animals undergoing parallel X-ray and chemical treatment were much stronger. This is illustrated by the curves shown in Fig.15. Edathamil calcium disodium was chosen in this case as an inhibitor. Here again, relations similar to those discussed in 3.2.2 could be observed: changes in radiation response were more rapid than those in melanin content of cells, radiosensitization preceding depigmentation. Tumour tissues with a marked melanization revealed a lower radioresistance, as manifested by a smaller size of tumours (see Fig.15), sometimes leading to their regression.

Parallel series of control experiments demonstrated [41] that a similar radiosensitization cannot be obtained in the amelanotic form of Bomirski melanoma, the effect appearing to be specific to pigment-containing cells.

4. SOME FUTURE PROSPECTS

The data discussed in Section 3 substantiate the hope that some effects useful for radiotherapy can be produced by influencing the normal activity of the melanin system, in the sense of its enhancement or inhibition. In the first case, improvement of radioresistivity should be expected, and in the second, a reverse change, i.e. sensitization should take place.

The significance of a possible protection, however, seems to be limited to radioresistance of the skin. Darkly pigmented skin is usually somewhat less susceptible to both u.v. and ionizing radiations, but the role of this factor does not seem to have been systematically studied as regards the latter type of radiation. On the other hand, the radiation tolerance of the skin is often a critical question for radiotherapists. Thus, it might perhaps be useful to define in more exact terms the extent to which the efficiency of a natural protection due to the whole (chemically activated) pigment system may help in counteracting radiation damage.

Certainly much more important is another aspect, i.e. radiosensitization in view of its possible applications in the therapy of human melanoma. Some clinical trials along this line are in progress. It is too early at the present stage to draw definite conclusions. Nevertheless, the results obtained so far are highly encouraging: cases of a very rapid regression were noted after combined radiation and chemical treatment, and no growth renewal took place over a period of 6–11 months. Twelve patients are under current investigation to establish whether the destruction of the tumour will prove irreversible.

Much work must still be done before the formulation of indications for wider use in clinical practice. There is no doubt, however, that further efforts in this direction are warranted.

The author is convinced that the question as to whether or not radiosensitization by the interference with the melanin system is possible, can be answered in the affirmative. Therefore, the main problems to be faced now are rather: to discover the ways of producing the most effective sensitization, and to identify in full detail the molecular mechanisms underlying the observed experimental facts.

REFERENCES

[1] BRUCE, A.K., Extraction of the radioresistant factor of *Micrococcus radiodurans*, Radiat. Res. 22 (1964) 155.
[2] BRUCE, A.K., MALCHMAN, W.M., Radiation sensitization of *Micrococcus radiodurans, Sarcina lutea, E. coli*, by p-hydroxymercuribenzoate, Radiat. Res. 24 (1965) 473.
[3] RUBAN, E.L., LIAKH, S.P., Investigation of natural melanins (in Russian), Izv. Akad. Nauk SSSR, Ser. Biol. 3 (1968) 530.
[4] COOPER, M., MISHIMA, Y., "The radio-response of malignant melanomas pretreated with chlorpromazine", Biology of Normal and Abnormal Melanocytes (KAWAMURA, T., FITZPATRICK, T.B., SEIJI, M., Eds), University of Tokyo Press, Tokyo (1971).
[5] SWARTZ, H.M., "Cells and tissues", Ch.4, Biological Applications of Electron Spin Resonance (SWARTZ, H.M., BOLTON, J.R., BORG, D.C., Eds), Wiley, New York (1972).
[6] ŁUKIEWICZ, S., The biological role of melanin. I. New concepts and methodical approaches, Folia Histochem. Cytochem. 10 1 (1972) 93.
[7] SARNA, T., SUBCZYNSKI, W., ŁUKIEWICZ, S., Interaction of melanins with spin labels (in Polish) (Proc. Vth Polish Conf. of Radiospectroscopy and Quantum Electronics), Poznań (1972) 45.
[8] LONGUET-HIGGINS, H.C., On the origin of free radical property of melanin, Arch. Biochem. Biophys. 86 (1960) 231.
[9] PULLMAN, A., PULLMAN, B., The band structure of melanins, Biochim. Biophys. Acta 66 (1963) 164.
[10] PULLMAN, B., PULLMAN, A., Quantum Biochemistry, Interscience Publ., New York (1963).
[11] COMMONER, B., TOWNSEND, J., PAKE, G., Free radicals in biological materials, Nature, London 174 (1954) 689.
[12] MASON, H.S., INGRAM, D.J.E., ALLEN, B., The free radical property of melanins, Arch. Biochem. Biophys. 85 (1960) 225.

[13] DAIN, A., KERKUT, G.A., SMITH, R.C., MUNDAY, K.A., WILMHURST, T.H., The interaction of free radicals in protein and melanin, Experientia 20 (1964) 76.
[14] PRASAD, K.N., JOHNSON, H.A., Differential radiosensitivity of the hepatic melanocytes of *Amphiuma tridactylum*, Radiat. Res. 33 (1968) 403.
[15] ZDANOVA, N.N., POKHODENKO, V.D., EPR spectra and radioresistance of some species of the *Dematiaceae* family (in Russian) Izv. Akad. Nauk SSSR, Ser. Biol. (1970) 83.
[16] NICOLAUS, R.A., Melanins (LEDERER, E., Ed.), Hermann, Paris (1968).
[17] GORDON, M., The Biology of Melanomas (GORDON, M., Ed.), The New York Academy of Sciences, New York (1948).
[18] DELLA PORTA, G., MÜHLBOCK, O.(Eds), Structure and Control of the Melanocyte, Springer, New York (196
[19] RILEY, V. (Ed.), Pigmentation: Its Genesis and Biological Control, Appleton-Century-Crofts, New York (1972)
[20] RILEY, V. (Ed.), Mechanisms in Pigmentation, S. Karger, Basel (1973).
[21] THATHACHARI, Y.T., Physical studies on melanins, J.Sci. Ind. Res. 30 (1972) 529.
[22] COPE, F.W., SEVER, R.J., POLIS, B.D., Reversible free radical generation in the melanin granules of the eye by visible light, Arch. Biochem. Biophys. 100 (1963) 171.
[23] ABLEWICZ-STRZAŁKOWSKA, E., RESZKA, K., ŁUKIEWICZ, S., "EPR signals of a model system: polymer-radioprotector in the X-ray field" (in Polish) (Proc. Polish Conf. on Radio- and Microwave Spectroscopy), Poznań (1975) 485.
[24] RUBAN, E.L., LIAKH, S.P., The stability of *Nadsoniella nigra* to the action of short-wave electromagnetic irradiation (in Russian), Izv. Akad. Nauk SSSR, Ser. Biol. (1968) 402.
[25] COBB, J.P., Effect of in vitro X-irradiation of pigmented and pale slices of Cloudman S91 mouse melanoma as measured by subsequent proliferation in vivo, J. Natl. Canc. Inst. 17 (1956) 657.
[26] GOLBERT, Z.V., LARIOSHCHENKO, T.G., PAPLIAN, N.P., Radiosensitivity of malignant melanoma (in Russian) Trudy Kazachskovo Instituta Onkologii i Radiologii 6 (1969) 100.
[27] DEWEY, D.L., The radiosensitivity of melanoma cells in culture, Br. J. Radiol. 44 (1971) 816.
[28] VINOGRADOVA, V.G., SHELIMOV, B.N., FOK, N.W., Khim. Vys Ehnerg. 2 (1968) 128.
[29] VINOGRADOVA, V.G., SHELIMOV, B.N., FOK, N.W., Khim. Vys Ehnerg. 2 (1968) 136.
[30] UNGAR, I.S., GAGER, W.B., LEININGER, R.I., J. Polymer Sci. 44 (1960) 296.
[31] IWASIOW, B., KAPISZEWSKA, M., "Comparative studies on the radioresistance of Bomirski hamster melanoma in situ and in vitro" (Proc. IVth Symp. Polish Soc. Medical Physics), Kazimierz Dolny (1975) 16.
[32] KAPISZEWSKA, M., Personal communication (to be published).
[33] CIESZKA, K., FOLWARCZNA, H., Personal communication.
[34] HAMADA, T., MISHIMA, Y., Intracellular localisation of tyrosinase inhibitor in amelanotic and melanotic malignant melanoma, Br. J. Dermatol. 86 (1972) 385.
[35] FUNAN HU, "Melanin production in mammalian cell culture", Biology of Normal and Abnormal Melanocytes (KAWAMURA, T., FITZPATRICK, T.B., SEIJI, M., Eds), University of Tokyo Press, Tokyo (1971).
[36] FORREST, I.S., GUTMANN, F., KEYZER, H., In vitro interaction of chlorpromazine and melanin, Aggressiologie 7 (1966) 147.
[37] LEWANDOWSKI, W., BIELEC, M., ŁUKIEWICZ, S., Chemically induced radiosensitization of *Aspergillus ornatus* mycelia, Folia Histochem. Cytochem. (to be published).
[38] PAJAK, S., SARNA, T., ŁUKIEWICZ, S., "ERP studies on the depigmentation of *Xenopus laevis* eggs" (in Polish) (Proc. Polish Conf. on Radio- and Microwave Spectroscopy), Poznań (1975) 503.
[39] DUMONT, J.N., EPPIG, J.J., Jr., A method for the production of pigmentless eggs in *Xenopus laevis*, J.Exp. Zool. 178 3 (1971) 307.
[40] HARPULA, Z., RESZKA, K., ŁUKIEWICZ, S., "The action of phenylthiourea on the paramagnetic properties and radiosensitivity of *Xenopus laevis* embryos" (in Polish)(Proc. Polish Conf. of Radio- and Microwave Spectroscopy), Poznań (1975) 507.
[41] ŁUKIEWICZ, S., IWASIOW, B., PAJAK, S., ABLEWICZ, E., "Chemically induced modifications of radiosensitivity in hamster melanoma" (in Polish) (Proc. IVth Symp. Polish Soc. Medical Physics), Kazimierz Dolny (1975) 18.
[42] SWARTZ, H.M., ŁUKIEWICZ, S., THOMM, G., Unpublished data.

MPG (2-MERCAPTOPROPIONYLGLYCINE)
A review on its protective action against ionizing radiations

T. SUGAHARA
Faculty of Medicine,
Kyoto University,
Kyoto, Japan
and
P.N. SRIVASTAVA
School of Life Sciences,
Jawaharlal Nehru University,
New Delhi, India

Abstract

MPG (2-MERCAPTOPROPIONYLGLYCINE): A REVIEW ON ITS PROTECTIVE ACTION AGAINST IONIZING RADIATIONS.

Many sulphydryl compounds have been reported to protect animals against lethal doses of ionizing radiations although such protection may not be universal. Their clinical application, however, has been very much limited by their high toxicity. It has been shown that 2-mercaptopropionylglycine (MPG) is a radioprotector effective in a very low dose (20 mg/kg) far below its toxic dose (2100 mg/kg) in mice. The compound is commercially available and used clinically as a detoxicating agent with a wide range of applications in many countries. Experimental studies on the protective effect of MPG and its related compounds are reviewed. At present it may be concluded that (i) MPG is the best protector amongst its own related compounds so far tested; (ii) from the standpoint of dose-effect relationship and toxicity, MPG and some of its related compounds seem to be quite different from cysteamine and its related compounds; (iii) the DRF of MPG is 1.3—1.4 for bone-marrow death in mice which is less than that of WR-2721; and (iv) MPG is the only chemical available for man at present in the dose range comparable to the dose for protection in mouse. For clinical application in improving tumour radiotherapy, approaches from both sides may be desirable such as the use of MPG, which is better from the toxicity standpoint, and the use of WR-2721, which is better for effectiveness.

INTRODUCTION

Ever since Patt and his associates [1] observed the radioprotective effect of cysteamine in animals, numerous compounds have been tested for their ability to protect against ionizing radiations. Many sulphydryl compounds have been reported to protect animals against lethal doses of radiations although such protection may not be universal. Their clinical application has been very much limited because of their high toxicity. The present paper reviews the experimental studies on the protective effect of 2-mercaptopropionylglycine (MPG) which protects against ionizing radiations at a very low dose (20 mg/kg) in mice. This dose is far below its toxic dose (2100 mg/kg). MPG is already being used clinically as a detoxicating agent in various countries with a wide range of application.

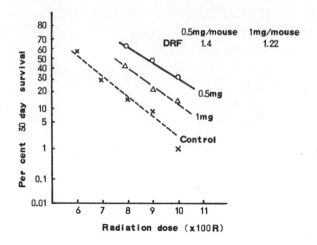

FIG.1. Dose-survival relationship of mice given MPG 15 min before irradiation, MPG doses of 0.5 mg/mouse or 1.0 mg/mouse were administered intraperitoneally. (From Sugahara et al. [2].)

STUDIES IN MICE

Three series of experiments have been carried out in mice to study the protective effects of MPG.

In one series, MPG was injected into mice intraperitoneally in a dose of 0.5 mg/animal (which will be equivalent to roughly 20 mg/kg body weight) 15 min before irradiation (1000 R). The per cent survival on day 30 (more than 30%) was higher than that of the controls (less than 1%). At this dose of MPG the $LD_{50/30}$ of treated animals is 880 R as compared to 620 R for the controls. The dose reduction factor (DRF), therefore, comes to 1.4. The DRF is reduced to 1.22 when the dose of MPG is increased to 1.0 mg/animal, i.e. approximately 40 mg/kg body weight. It is thus clear that MPG is a better protector at a lower dose (20 mg/kg) than at a higher dose (40 mg/kg) [2, 3] (see Fig.1).

The histopathological and histochemical effects of MPG have been screened at various low doses. No adverse effect (either qualitative or quantitative) could be detected in the tissues examined (testis, ovary, liver, spleen, bone-marrow, adrenal and intestine) for three weeks at 20 mg/kg body weight. A slight effect was observed only at doses above 200 mg/kg body weight [4].

In the second series of experiments, various derivatives of MPG were screened for their radio-protective activity. Fifteen derivatives were tested in lethally irradiated mice. Amongst the derivatives, only 3-MPG and MPG-amide showed some protection, but none of them showed better protection than MPG. Between the two, however, MPG-amide was more effective [5].

In the third series, MPG was compared with cysteamine in the doses of 20 and 200 mg/kg. MPG has a comparable protective action to cysteamine at the dose of 20 mg/kg although the safety margin of MPG is larger, since this dose is approximately 1.0% of the toxic dose of MPG and 7.3% of the toxic dose of cysteamine. Strangely enough, in this experiment the per cent survival at 30 days after MPG treatment was found to be 58.8 which is higher than what had been achieved earlier (see Table I).

Urano and Tsukiyama [6] have recently used MPG in mammary carcinoma in C3H/He mice to study its radioprotective effect. The tumour was locally irradiated with 4000 R of X-rays with or without MPG (20 mg/kg) pretreatment under normal (air) and hypoxic (by local

TABLE I. PER CENT SURVIVAL OF MICE IRRADIATED WITH 1000 R OF COBALT-60 GAMMA RAYS AND TREATED WITH MPG OR CYSTEAMINE

Chemicals	Dose (mg/kg)	% of toxic LD_{50}	% survival for 30 days [a]
MPG	20	0.9	58.8
	200	9.0	76.6
Cysteamine	20	7.3	53.3
	200	72.7	84.0
Physiological saline	0.1 ml/mouse		0.0

[a] Groups of 20 mice were observed twice for each treatment.

TABLE II. TUMOUR REGROWTH TIME (TRT) FOR C3H/He MICE MAMMARY CARCINOMA AFTER TREATMENT WITH MPG

	X-ray dose	X-ray condition	MPG (mg/kg)	$TRT_{(50)}$ (days)	(95% Conf. limits)	DMF
A	4000	Air	0	23.4	(20.4–24.0)	
	4000	Air	20	18.8	(17.6–20.0)	1.19
B	4000	Hypoxia	0	20.3	(19.3–21.5)	
	4000	Hypoxia	20	18.7	(18.0–19.5)	1.09
C	800 × 5	Air	0	18.6	(18.1–19.2)	
	800 × 5	Air	20	19.1	(18.1–20.2)	0.97

application of a heavy brass clamp above the tumour) conditions. As shown in Table II, the dose modifying factor (DMF) in air was 1.19. In normal mice, a dose reduction factor of 1.4 is achieved after using 20 mg/kg body weight of MPG. Under hypoxia or fractionated irradiation no significant protection was observed.

STUDIES IN MAN

Clinical observations on the radioprotective action of MPG have been made by Tanaka [7] in cervical carcinoma patients receiving routine radiotherapy. He has observed that 250 mg/patient in 20% glucose given 15–30 min before irradiation was enough to produce radioprotection. Higher doses did not increase the protective effect. The leucocyte counts, especially the per cent lymphocyte count, recovered earlier in the MPG-treated patients than in the non-treated group. When radiation was given beyond 2000 R, the chromosomal aberration in the MPG-treated group was remarkably lower. MPG also prevented radiation sickness. The radioprotective effect of MPG was observed in leucocyte counts by double blind test as well (see Figs 2–5).

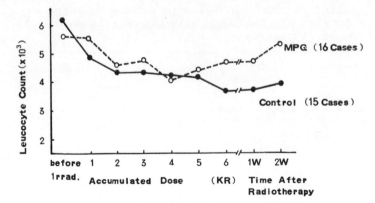

FIG.2. Changes in leucocyte counts of the control and MPG-treated groups during and after radiotherapy in man. (From Tanaka [7].)

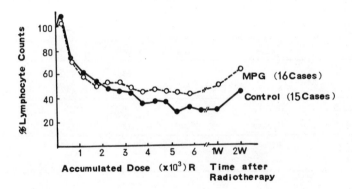

FIG.3. Changes in % lymphocyte counts of the control and MPG-treated groups during and after radiotherapy in man. (From Tanaka [7].)

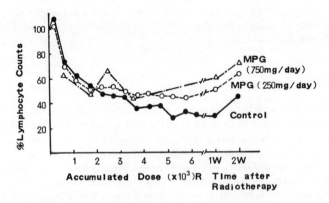

FIG.4. Changes in % lymphocyte counts in the patients treated with 250 mg and 750 mg MPG. (From Tanaka [7].)

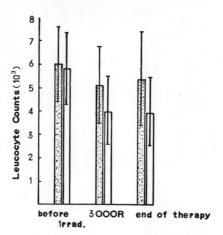

FIG.5. Comparison of lymphocyte counts in control and MPG-treated groups after 3000 R and at end of treatment by double blind test. (From Tanaka [7].)

TABLE III. RADIATION PROTECTION IN CULTURED MAMMALIAN CELLS AFTER MPG TREATMENT

Cell line	Pre-treatment		$D_{(0)}$	Extrapolation number	Ref.
	Concentration	Time (min)			
Chinese hamster	0.1 mg/ml	30	90	2.13	8
	None		88	2.57	
Mouse L	0.005 mg/ml	15	170	6.0	9
	None		120	9.0	
	0.02 mg/ml	15	160	6.0	
	None		128	9.0	
BW5147 Ascites tumour	1.0/mouse	15—20	260	2.3	10
	None		240	1.5	

TISSUE CULTURE STUDIES

Colony formation

Tissue culture studies are valuable not only for screening radioprotective drugs but also for elucidating the mechanism of chemical radiation protection. Révész et al. [8] reported some data obtained in collaboration with Littbrand that treatment of Chinese hamster cells with MPG for 30 min before irradiation does not affect the slope of radiation survival curves either under oxic or anoxic conditions. But the MPG treatment does raise significantly the extrapolation number of the anoxic survival curve. On the other hand, the extrapolation number of the survival curve of cells irradiated under oxic conditions remains unchanged.

TABLE IV. CHEMICAL RADIATION PROTECTION IN MOUSE AND MAN

Compound	MOUSE				MAN		
	Biological features	Toxicity LD_{50} (mg/kg)	Protective dose	DRF based on $LD_{50/30}$	Administered dose (mg)	Protection against:	
						Leucopenia	Chromosome aberration
Cysteine	Natural	1500	1200 i.v.	1.42		No trial	
Glutathione	Natural	4000	4000 i.p.	1.28	200	Yes	—
			1600 i.p.	1.12			
Cysteamine	Toxic	275	150 i.p.	1.45	200–400	Yes	—
AET	Toxic	690	400 i.p.	2.15	100–200	Yes	—
WR-2721	Toxic	A/J 554	80 i.p.	1.47		No trial	
			400 i.p.	2.78			
		BALB/cJ 784	56 i.p.	1.35			
			120 i.p.	1.55			
			400 i.p.	2.44			
MPG	Non-toxic	2100	20 i.p.	1.40	250	Yes	Yes

Horikawa and Hikita [9] have recently studied the effect of MPG on mouse L cells in culture at various dose levels and observed some protection with very low concentrations of MPG. The latest personal communication from U [10] shows that MPG has a positive radiation protection effect in cultured mammalian cells. BW5147 tumour cells were irradiated in vivo and assayed in vitro. The mice weighed between 19 and 24 g and were given MPG in a dose of 1 mg/mouse, 15—20 min before irradiation. These findings are summarized in Table III.

Mitotic index

Kawasaki et al. [11—13] have been actively engaged in studying the radioprotective effect of SH-compounds basing their observations on radiation-induced mitotic delay in mouse cultured L cells. According to them the radiation protection is of two types. In one the mitosis is delayed for a long time while in the other there is only a short mitotic delay followed by a rapid rise of mitotic index. The former is observed when cysteamine is used and the latter when glutathione is used. Their latest personal communication indicates that MPG falls into the second category but its radiation protection effect is greater than that of GSH in the same molar concentration.

DISCUSSION

Radiation protection, dose and toxicity

Table IV summarizes the effectiveness of various sulphydryl compounds in mouse and man as reported in literature together with their DRF values. Cysteamine and, more recently, WR-2721 have been reported as the most potent drugs for radiation protection [14—17]. However, they have not yet been used clinically because of their high toxicity. For a DRF of 1.4, roughly 50% of the toxic dose of cysteamine and 14% of that of WR-2721 have to be used which are considerably high. On the other hand, only 1% of the toxic dose of MPG has to be used to achieve the same DRF. This dose of MPG does not cause any histopathological damage and the chemical is already being used clinically as a detoxicating agent in many countries.

It may be relevant to compare the histopathological damage caused by serotonin to various tissues. It has been shown that 75 mg/kg of serotonin (the dose required to achieve a DRF of 1.4) causes severe pathological damage in gonads, liver, adrenal and thyroid in mice [18]. Tricou and Doull [19] observed the best radioprotective effect of serotonin in mice after a dose of 50-100 mg/kg with an optimum at 90 mg/kg which gives a DRF of 1.6. This is roughly 10% of the toxic LD_{50} dose. It would be worth while examining the histopathological effect of cysteamine and WR-2721 at the dose levels at which they bring the DRF to 1.4 for which 50% of the toxic dose of cysteamine and 14% of WR-2721 would have to be used.

Sigdestad et al. [20] have used two-thirds of the toxic $LD_{50/6}$ dose of WR-2721 in mice, but this dose did not modify the inherent radiosensitivity of the intestinal crypts although the DRF came to 1.64. However, Harris and Phillips [16] have shown that WR-2721 and WR-638 protect euoxic marrow-colony-forming units in vivo better than does cysteamine when the cells are treated with both drug and X-radiation before transplantation. Under these conditions, the DMF for WR-2721 is 3.0 and for WR-638, 2.3 which can be compared with the DMF of 1.7 for cysteamine [21], 2.3—2.6 for cystamine [22], and 1.7—2.2 for AET [23, 24]. However, the dose of thiophosphates used by Sigdestad et al. [20] was 600 mg/kg when the toxic LD_{50} dose was 785 mg/kg. These authors also warned that the toxicity of the thiophosphates will undoubtedly determine whether or not they gain clinical acceptance. Large quantities of intracellular free SH, such as the one present after administration of aminothiols [25, 26, 27], are probably responsible for the toxicity because of the drastic alterations which they cause in cellular redox potential and because of oxidation to disulphides, which irreversibly poison essential enzymes [28]. The phosphate group preventing oxidation of the SH function may limit toxicity. On the other hand, the redox potential of MPG is naturally highest and it decreases in the following order: glutathione, cysteamine and cysteine which may in itself explain their toxicity. The high redox potential of MPG may be related to the radiation protection quality of this drug [29]. Révész et al. [8] have also shown that MPG releases less glutathione, which may be toxic to cells, than cysteamine at comparable concentrations.

Classification of chemical protectors

There are basically two types of sulphydryl protectors reported so far: one is of natural origin with low toxicity such as glutathione and the other is artificial with high toxicity such as cysteamine (including WR-2721 which is a derivative of cysteamine). Both of them, especially glutathione, should be used in very large doses to achieve successful protection. MPG appears to be of the third type which is artificially synthetic with a very low toxicity and, at the same time, effective as a radioprotector at a very low dose. This may indicate that MPG is not only useful for practical application but also for elucidating the mechanism of chemical radiation protection.

Hikita et al. [30] have shown that sulphydryl compounds could be put under three classes based on their radioprotective action. They had taken the colony-formation ability of mouse L cells as the criterion. The most effective radioprotection was obtained by cysteamine followed by AET and MPG-amide (second category). It has recently been shown by them that glutathione will also belong to this latter class. MPG in concentrations around 0.02 mM and 15 mM (the third category) has shown a low but significant protective action. The most important thing to note about the third category is that it does not show any toxic effect, whereas toxicity had been observed in cysteamine and cysteine in the concentration range of 0.1–2.0 mM. It is rather strange that cysteamine and cysteine, although toxic in the concentration range of 0.1–2.0 mM, are less toxic in higher ranges. Similar paradoxical dose-dependence has been observed by Delrez and Firket [31] in Chinese hamster cells, by Sawada and Okada [32] in mouse leukemic cells (L 5178Y) and by Takagi et al. [33] in HeLa cells after treatment with cysteamine. The real mechanism is not yet clear although cell-killing action has been ascribed to the generation of peroxide ($2 RSH + O_2 = RSSR + H_2O_2$) by the compound.

Evaluation in clinical application

It is difficult to conduct experiments in man in the same way as in mice. The only available cases of human exposure, therefore, are from cancer patients undergoing radiotherapy. The irradiation conditions in such cases are local and fractionated which are different from whole-body irradiations of mice. The results obtained from such cancer patients should be evaluated critically. In this respect, the chromosome aberration could be a reliable index of protection [34], though the patterns of aberration yield may be modified by various factors [35, 36]. Great care, however, has to be exercised even for such a dosimetric method. The patterns may appear to be simple as there is a direct relationship between dose and response, but in reality they are not as simple as they look. Furthermore, the dose-rate or, in fractionated irradiation, the time interval between sessions can cause variations in the yield of chromosome aberration. The yield of chromosome aberration in lymphocytes is also influenced by physiological factors such as age, cell kinetics, location of lymphocytes at the time of irradiation etc. When we consider the number of factors involved and their complexity, the agreement between observed dose measurement and evaluation based on chromosome aberration seems very good, but it is clear that further development of a practical method necessitates a large amount of basic research.

Sufficient data are not available for the protection and toxicity of MPG and WR-2721 to be compared at the same dose levels or at the same DRF level. Further work is necessary in this direction. However, for clinical application, improvements in tumour radiotherapy may have to be sought by approaching the problem from both sides, namely by using the less toxic MPG or the more effective WR-2721.

The properties of chemical protectors essential for radiotherapy have been discussed before [16, 37]. A chemical protective agent which would lower the sensitivity of oxygenated tumour cells to the level of anoxic cells would eliminate the differential sensitivity and permit the use of doses large enough to effectively kill a large fraction of the latter. We do not know if such a "therapeutic nirvana" (as referred to by Harris and Phillips [16]) is available yet. Elimination of the two- to three-fold difference of anoxic and eouxic cells has long been one of the major objectives of the radiotherapist.

Differential protection of the tumour and adjacent normal tissue is the most important among various factors. A few papers have been published on WR-2721 in which DRFs for normal tissue, such as bone-marrow death, and tumour cells have been compared [16, 38–43]. Preliminary data on MPG by Urano and Tsukiyama [6] seem to support the differential protection as well. A model system for screening chemical protectors useful for radiotherapy should be developed.

REFERENCES

[1] PATT, H. M., TYREE, E. B., STRAUBE, R. L., SMITH, D. E., Science 110 (1949) 213.

[2] SUGAHARA, T., TANAKA, Y., NAGATA, H., TANAKA, T., KANO, E., "Proceedings of the International Symposium on Thiola". Santen Pharmaceutical Co. Ltd., Osaka, Japan (1970) 267.

[3] NAGATA, H., SUGAHARA, T., TANAKA, T., J. Radiat. Res. 13 (1972) 163.

[4] SRIVASTAVA, P. N., Unpublished.

[5] SUGAHARA, T., NAGATA, H., HORIKAWA, M., HIKITA, M., Experientia (in press).

[6] URANO, N., TSUKIYAMA, I., Personal communication.

[7] TANAKA, Y., "Proceedings of the Second International Symposium on Thiola". Santen Pharmaceutical Co. Ltd., Osaka, Japan (1972) 23.

[8] REVESZ, L., MODIG, H., MONSTANTINOVA, M. M., Ibid (12).

[9] HORIKAWA, M., HIKITA, M., Personal communication.

[10] U, RAYMOND, Personal communication.

[11] KAWASAKI, S., KOBAYASHI, M., OKITA, I., SUETOMI, K., SAKURAI, K., Nippon Acta Radiologica 34 (1974) 444.

[12] KAWASAKI, S., KOBAYASHI, M., OKITA, I., SAKURAI, K., Nippon Acta Radiologica 34 (1974) 599.

[13] KAWASAKI, S., KOBAYASHI, M., OKITA, I., SUETOMI, K., SAKURAI, K., Nippon Acta Radiologica 34 (1974) 599.

[14] YUHAS, J. M., Radiat. Res. 44 (1970) 621.

[15] YUHAS, J. M., Radiat. Res. 47 (1971) 526.

[16] HARRIS, J. W., PHILLIPS, T. L., Radiat. Res. 46 (1971) 362.

[17] YUHAS, J. M., PROCTOR, J. O., SMITH, L. H., Radiat. Res. 54 (1973) 222.

[18] SRIVASTAVA, P. N., "Biological Aspects of Radiation Protection". (SUGAHARA, T., HUG, O., Eds.) Igaku Shoin Ltd., Tokyo (1971) 157.

[19] TRICOU, B. J., DOULL, J., University of Chicago USAF Radiation Laboratory Quarterly Progress Report No. 35 (1960) 70.

[20] SIGDESTAD, C. P., CONNOR, A. M., SCOTT, R. M., Radiat. Res. 62 (1975) 267.

[21] SMITH, W. W., BUDD, R. A., CORNFIELD, J., Radiat. Res. 27 (1966) 363.

[22] JRASKOVA, V., TRADLECEK, L., Radiat. Res. 30 (1967) 14.

[23] DUPLAN, J. F., FUHRER, J., C. R. Soc. Biol. 160 (1966) 1142.

[24] AINSWORTH, E. J., LARSEN, R. M., Radiat. Res. 40 (1969) 149.

[25] BALL, C. R., Biochem. Pharmacol. 15 (1966) 809.

[26] MODIG, H. G., REVESZ, L., Int. J. Radiat. Biol. 13 (1967) 469.

[27] SORBO, B., Arch. Biochem. Biophys. 98 (1962) 342.

[28] NESBAKKEN, R., ELDJARN, L., Biochem. J. 87 (1963) 526.

[29] SUGAHARA, T., "Proceedings of the Second International Symposium on Thiola". Santen Pharmaceutical Co. Ltd., Osaka, Japan (1972) 17.

[30] HIKITA, M., HORIKAWA, M., MORI, T., J. Radiat. Res. 16 (1975) 162.

[31] DELREZ, M., FIRKET, H., Biochem. Pharmacol. 17 (1968) 1893.

[32] SAWADA, S., OKADA, S., Radiat. Res. 44 (1970) 116.

[33] TAKAGI, Y., SHIKITA, M., TERASIMA, T., AKABOSHI, S., Radiat. Res. 60 (1974) 292.

[34] TANAKA, Y., SUGAHARA, T., J. Radiat. Res., 11 (1970) 166.

[35] TAMURA, H., SAKURAI, M., SUGAHARA, T., Blood 36 (1970) 43.

[36] TAMURA, H., SUGIYAMA, Y., SUGAHARA, T., Radiat. Res. 59 (1974) 653.

[37] SUGAHARA, T., "Fraction Size in Radiobiology and Radiotherapy". (SUGAHARA, T., REVESZ, L., SCOTT, O. C. A., Eds.) Igaku Shoin Ltd., Tokyo (1973) 84.

[38] YUHAS, J. M., STORER, J. B., J. Nat. Cancer Inst. 42 (1969) 331.

[39] YUHAS, J. M., J. Nat. Cancer Inst. 48 (1972) 1255.

[40] YUHAS, J. M., J. Nat. Cancer Inst. 50 (1973) 69.

[41] LOWY, R. O., BAKER, D. G., Acta Radiol. Ther. Phys. Biol. 12 (1973) 425.

[42] UTLEY, J. F., PHILLIPS, T. L., KANE, L. J., WAHARAM, M. D., WARA, W. M., Radiology 110 (1974) 213.

[43] PHILLIPS, T. L., KANE, L., UTLEY, J. F., Cancer 32 (1973) 528.

RADIOPROTECTORS AND
RADIOTHERAPY OF CANCER*

J.R. MAISIN, M. LAMBIET-COLLIER,
G. MATTELIN
Centre d'études nucléaires/Studiecentrum voor Kernenergie,
Mol, Belgium

Abstract

RADIOPROTECTORS AND RADIOTHERAPY OF CANCER.
The administration of mixtures of radioprotectors not only increases the degree of protection for the 30-day survival compared with that of AET alone but also exerts a favourable influence on the life shortening of mice exposed to single or repeated total body X-irradiation. Although sulphydryl radioprotectors penetrate in the cancer tissues and protect them against radiation damage, AET and mixtures of chemical protectors were found to be useful in the radiotherapy of tumours induced by grafting Landschutz cells in the leg or in the peritoneal cavity of BALB/c mice under certain conditions.

INTRODUCTION

Radioprotective compounds were used in experimental radiotherapy of cancer [1–12] already soon after their discovery in 1949. However, it was soon recognized that, unfortunately, their toxicity, their low protective effects and the lack of a selective protection of normal tissues compared with cancer tissues rendered the general use of radioprotectors in human radiotherapy of only doubtful value [10–12].

This paper presents a synopsis of various investigations that we have carried out during the past years. It deals with:

(a) The radioprotective action of 2-β-aminoethylisothiuronium-Br-HBr (AET) or of a mixture of radioprotectors against short and long-term survival;
(b) The localization and distribution of ^{3}H-AET in normal and cancer cells,
(c) The use of AET and of a mixture of radioprotectors in the experimental radiotherapy of Landschutz tumours implanted in the leg or in the peritoneal cavity of BALB/c mice.

MATERIAL AND METHODS

A detailed description of the different techniques used has been published earlier [11–18]. In summary, BALB/c male mice, 12 weeks old, weighing 25 to 30 g (50/30 days LD, 576 R) were used for all survival experiments. The conditions of irradiation were 300 kV, 20 mA, 1.5 mm Cu, and an exposure rate of 100 R/min. In most of the experiments, the mice were exposed to a single dose of X-irradiation. The time of administration and the dose of the chemical compounds used in the experiments are given in Table I. The mice were housed two

---

* This work was supported by Euratom/CEN contract No. 095-72-1 BIO B.

TABLE I. TIME OF ADMINISTRATION AND DOSES OF CHEMICAL COMPOUNDS USED

Treatment	Amount of compound (mg)	Route of administration	Time of treatment (minutes before irradiation)
AET[a]	8	i.p.	10
AET[c]	5	i.p.	10
GSH[b]	16	orally	25
5-HT[c]	1	i.p.	5
Cysteine	15	i.p.	15
MEA[d]	2.5	i.p.	10

[a] 2-β-aminoethylisothiuronium-Br-HBr
[b] Glutathione
[c] Serotonin creatine sulphate (5-hydroxytryptomine)
[d] Mercaptoethylamine

in a cage and given free access to food and water after irradiation. All mice that survived more than 30 days after an X-ray exposure were selected for the long-term studies. For some experiments, repeated doses of radiation were administrated at intervals of 15, 30 or 60 days. Protectors were given before each X-ray exposure. Because the chemical protectors are very toxic, especially when administered repeatedly, the amounts of the chemical protectors in the mixture were diminished starting from the second irradiation exposure [14, 15].

In the study on the distribution and localization of AET in the cells, tritium-labelled material synthetized by Dr. J.J. Dougherty of the Oak Ridge National Laboratory, Oak Ridge, Tennessee, by the Wilzbach procedure [13] was used.

Labelling was restricted to the hydrogen atoms of the methylene carbon. A 5.8-mg dose of AET, specific activity 0.3 µCi/mM, was injected intraperitoneally into normal and tumour-bearing mice.

The mice were killed at different times after injection of the compound. The tissues were fixed in neutral formalin. Slides of tissue 5 µm thick were exposed to a Kodak NTB2 emulsion. After exposure, the tissues were stained with Harris haematoxylin. The mean number of grains in the nuclei of various cancer and normal tissues was counted.

In our experiments on the influence of radioprotectors on the radiotherapy of cancer we used Landschutz ascites tumour cells. This tumour, described for the first time in 1954 by Tjio and Levan, has 46 chromosomes [11]. It was introduced in our laboratory in 1959 and maintained by transplantation in the peritoneal cavity of BALB/c mice [19].

BALB/c mice were injected in the peritoneal cavity or in one leg with different quantities of ascites cells and irradiated under different conditions (single or repeated doses given at 7-day intervals) (a) on the treated leg only, (b) on the leg and on the inferior part of the abdomen, and (c) on the abdomen for cells injected in the peritoneal cavity. The rest of the body was shielded with a plate of 4-mm-thick lead [10–12, 16, 17].

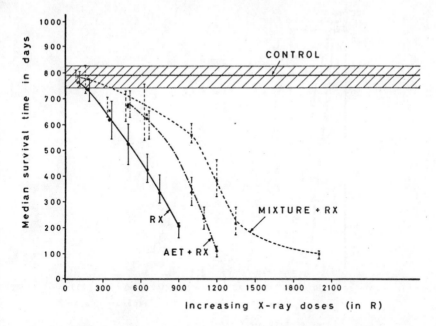

FIG.1. *Long-term survival of mice treated with AET or with a mixture of AET + 5-HT + GSH + cysteine + MEA.*

RESULTS

Acute survival

The dose reduction factors (DRF) obtained for the 30-day survival for mice treated before X-irradiation with AET or with a mixture of AET + GSH + 5 HT + cysteine + MEA were 1.7 and 2.8 respectively. If, in addition, three transplantations of isogenic nucleated bone-marrow cells were administered after irradiation, survival was improved still further (DRF = 3.7) [16].

Long-term survival

(a) Single dose of irradiation

The results obtained for the median survival time of BALB/c mice treated or not treated with chemical protectors before a single dose of radiation are summarized on Fig.1. For doses higher than 350 R the data for the non-protected mice approach a straight line indicating a linear relationship between life shortening and dose. For the protected irradiated mice, the dose-effect curves obtained for the different groups of mice are not parallel, a fact which is evident if one compares exposure with the low and the high dose range. The dose reduction factor obtained for the AET-treated mice was about 1.8 after a dose of 500 R and 1.3 after a dose of 1100 R of X-rays. Mice treated with a mixture of radioprotectors had about the same dose reduction factor of about 2 after exposure to 500 to 1200 R. After an exposure to 1350 R, the dose reduction factor fell to 1.5. For X-ray doses below 350 R, the differences between the mice protected with a mixture of radioprotectors and the irradiated non-protected mice is not significant.

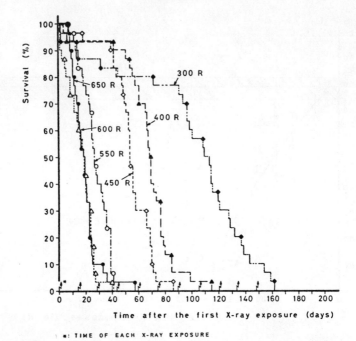

FIG.2a. Long-term survival of mice exposed to repeated X-ray exposures given at intervals of 15 days.

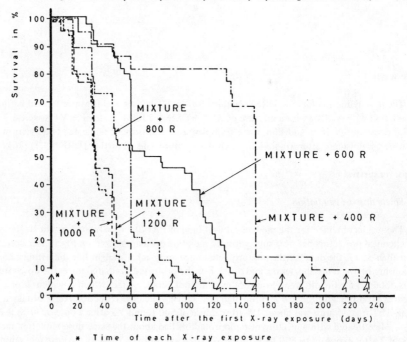

FIG.2b. Long-term survival of mice pre-treated with a mixture of radioprotectors before each X-ray exposure given at intervals of 15 days.

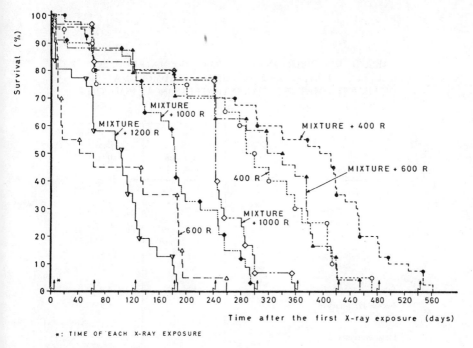

FIG.3. *Long-term survival of mice pre-treated with a mixture of radioprotectors before each X-ray exposure given at intervals of 60 days.*

(b) Repeated doses of irradiation

Survival after repeated doses of 400, 600, 800, 1000 or 1200 R at intervals of 15, 30 or 60 days was studied in mice protected with AET alone or with a mixture of radioprotectors. Non-protected mice exposed to doses of 400 or 600 R at an interval of 30 days died usually after a cumulative dose of 3600 and 1200 R respectively. All mice treated with AET and exposed to 600, 800, 1000 or 1200 R died after a cumulative X-ray dose of 4200, 4000, 3000 and 3600 R, whereas mice protected with a mixture of chemical compounds died only after cumulative doses of 6000, 6400, 6000 and 4800 R respectively.

Thus, the improvement in survival of mice treated with chemical radioprotectors for repeated X-ray exposure at 30-day intervals is higher than that obtained for short- and long-term survival after a single X-ray exposure. When a 15-day interval between exposures was chosen, the mice treated with a mixture of five radioprotectors supported about the same cumulative X-ray doses as those with the 30-day intervals (Figs 2a and 2b). On the other hand, the survival of protected mice exposed at intervals of 60 days is lower than that of those exposed at intervals of 15 or 30 days (Fig.3).

Many of the mice treated with mixtures of radioprotectors and irradiated with repeated X-ray doses at 30-day intervals died either shortly after treatment due to the toxicity of the mixture or later due to radiation pneumonitis. Yet, the haematopoietic and lymphopoietic tissues and the mucosa of the small intestine exhibited good regeneration. The spleen, the lymph nodes, the thymus and the Peyer's patches appeared almost normal on histological examination and the mucosa of the small intestine did not seem to be altered greatly, but the number of nuclei and mitoses in the crypts appeared increased compared with the controls.

TABLE II. DISTRIBUTION OF RADIOACTIVE MATERIAL IN NUCLEI OF SPONTANEOUS AND GRAFTED TUMOURS AT DIFFERENT TIMES AFTER AN INTRAPERITONEAL INJECTION OF ^{3}H-AET

	Time after injection (min)	Number of grains per 100 μm^2
Spontaneous tumour		
Adenocarcinoma of the ovary	30	6.4
Normal ovary		6.5
Mammary adenocarcinoma	30	10.3
Lymphosarcoma	30	4.9
Normal lymph node		5.9
Melanocarcinoma	60	7.7
Normal eye		7.3
Angioreticulosarcoma	60	14.9
Normal skin (epidermis)		2.7
Grafted tumour		
(Squamous cell carcinoma of the skin)		
Tumour	30	5.5
Epidermis		2.9
Tumour	50	1.8
Epidermis		1.2
Tumour	60	2.5
Epidermis		1.8

DISTRIBUTION AND LOCALIZATION OF AET IN NORMAL AND CANCEROUS TISSUES

Normal tissues

In brief, our experimental results indicate that:

1. Labelled AET penetrates into all tissues at a level proportional to their vascularization.
2. The highest radioactivity is obtained 10 min after the administration of AET; at this time most of the nuclei are labelled.
3. In a given tissue, the nuclei and the cytoplasm are labelled to the same degree.
4. Cells in mitosis are labelled.
5. Radioactivity is found on and close to the membranes of the nuclei, the cells and the tissues.

Cancer tissues

The results of Table II demonstrate that, after an intraperitoneal injection of ^{3}H-AET, the radioactivity in spontaneous and grafted cancers was at least the same or greater than that of the corresponding normal tissues or of the tissues surrounding the cancer. In highly vascularized tumours the radioactivity was even considerably higher than that of surrounding tissue. The intracellular localization of the grains in the tumours and healthy tissues was similar.

RADIOPROTECTORS AND THE TREATMENT OF CANCER

Graft of Landschutz cells in one leg

After a local irradiation of one leg with 3 times 2000 R 12 days after the graft of 62 500 ascites cells, the percentage of developing tumours is greater in protected mice than in mice irradiated without protection and, inversely, the survival of mice irradiated without protection is slightly better than the survival of protected ones (Figs 4a and 4b).

After an exposure to three doses of 3000 R, the take of the graft is about 50% in the two groups of irradiated mice (Figs 5a and 5b).

In a separate experiment, a suspension of about 33×10^6 ascites cells was injected into one leg of BALB/c male mice. Fifteen days after grafting the tumour cells, the mice were irradiated on the right leg and on the inferior part of the abdomen with 6000 R of X-rays given in three doses of 2000 R at 7-day intervals. The mice protected with AET received before the 1st, the 2nd and the 3rd irradiation an intraperitoneal injection of 8, 6 and 5 mg of AET respectively. The mice were killed at different time intervals and the tumour volume, the mitotic activity, the incorporation of ^{3}H-thymidine and the number of abnormal anaphases was determined.

1. The tumour volume of the protected mice was intermediate between that of the controls and that of the mice irradiated without protection.
2. In the non-protected irradiated mice, the mitotic activity in the tumour decreased markedly on the 1st day after each X-ray exposure and returned to normal on the 4th day. In the protected irradiated mice, the mitotic activity on the 1st and the 4th days was about the same as in the controls (see Table III).
3. The incorporation of ^{3}H-thymidine into the DNA on the first day after each exposure was markedly lower in non-protected mice than in protected irradiated mice and in control mice.
4. The number of abnormal anaphases was higher in the irradiated non-protected mice than in the two other groups (Table IV).

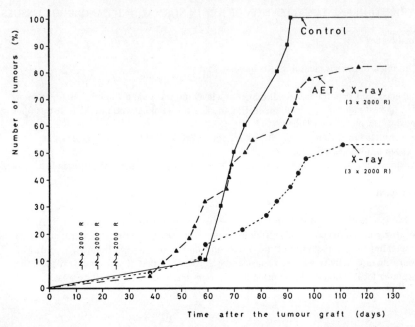

FIG.4a. Percentage of cancers in control mice and in mice irradiated with or without protection with 3 times 2000 R given at 7-day intervals (the first irradiation was given 12 days after the graft of 62 500 ascites cells in the leg).

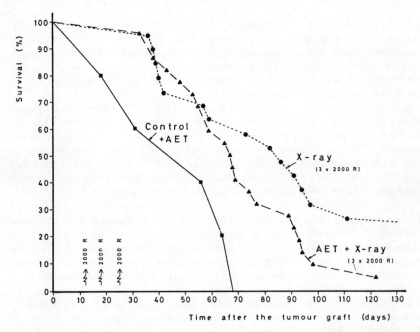

FIG.4b. Survival of BALB/c mice grafted in the right leg with 62 500 ascites cells of Landschutz and irradiated locally with or without protection. (The experimental conditions were the same as in Fig.4a.)

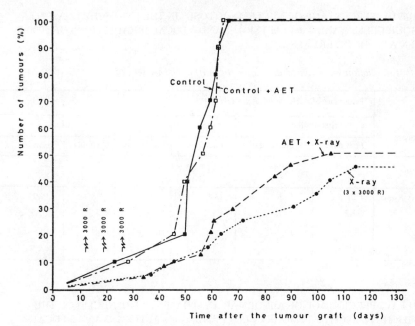

FIG.5a. Percentage of cancers in control mice and in mice irradiated with or without protection with 3 times 3000 R given at 7-day intervals (the first irradiation was given 12 days after the graft of 62 500 ascites cells in the leg).

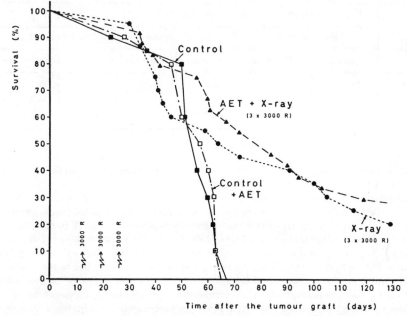

FIG.5b. Survival of BALB/c mice grafted in the right leg with 62 500 ascites cells of Landschutz and irradiated locally with or without protection. (The experimental conditions were the same as in Fig.5a.)

TABLE III. PERCENTAGE OF NUCLEI IN MITOSES IN THE LANDSCHUTZ ASCITES TUMOUR CELLS 1 AND 4 days AFTER 1, 2 AND 3 LOCAL IRRADIATIONS OF 2000 R GIVEN AT 7-day INTERVALS

The first irradiation was given 15 days after the graft of the ascites cells.

Treatment	Percentage of nuclei in mitoses after:					
	first irradiation		second irradiation		third irradiation	
	First day	Fourth day	First day	Fourth day	First day	Fourth day
Control	13.7 ± 1.9					
AET + X-ray	12.1 ± 1.2	13.6 ± 1.1	15.8 ± 1.2	13.1 ± 0.5	15.5 ± 3.5	15.2 ± 2.9
X-ray	6.7 ± 1.7	10.7 ± 0.4	7.8 ± 1.3	10.9 ± 0.4	6.3 ± 1.5	12.8 ± 2.7

TABLE IV. PERCENTAGE OF ABNORMAL ANAPHASES IN THE NUCLEI OF THE LANDSCHUTZ ASCITES TUMOUR CELLS 1 AND 4 days AFTER 1, 2 AND 3 LOCAL IRRADIATIONS OF 2000 R GIVEN AT 7-day INTERVALS

The first irradiation was given 15 days after the graft of the ascites cells.

Treatment	Percentage of abnormal anaphases in the nuclei after:					
	first irradiation		second irradiation		third irradiation	
	First day	Fourth day	First day	Fourth day	First day	Fourth day
Control	20	20.9	13.0	15	20	19.2
AET + X-ray	50	25.5	28	34.5	21.4	30.7
X-ray	66.6	25.6	86.2	61.5	40	21.0

In most of the control mice, metastatic lymph nodes were present already 36 days after the graft. Metastases appeared later in the two other groups.

Although chemical protectors thus also protected cancer cells as well as normal cells they may nevertheless be useful in the radiotherapy of cancer because the protected mice supported better the exposure to X-irradiation than the irradiated non-protected mice. If, in addition to the local irradiation of the tumour with 3 times 2000 R, a 2 cm^2 area of the abdomen was also exposed to 3 times 1500 R, the survival of the protected mice was much better than that of the control and of the irradiated non-protected mice in spite of the local protection of the tumour (Fig.6). This field of irradiation included intestinal loops, a part of the colon, of the liver and of the right kidney; the second and the fourth irradiation were followed by a transplantation of 1 million nucleated bone-marrow cells. At autopsy, the haematopoietic organs (bone marrow and spleen) and the inferior part of the colon which was in the irradiated field were less damaged in the protected mice than in the non-protected mice. In another experiment, mice were grafted

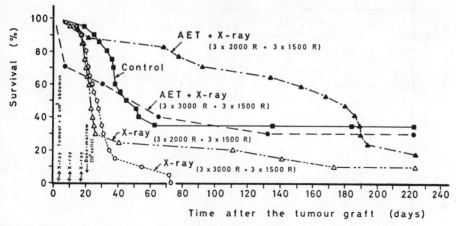

FIG.6. *Survival of BALB/c mice grafted with 62 500 Landschutz ascites tumour cells in the right leg and irradiated on the leg with 3 times 2000 R and on an area of 2 cm² of the abdomen with 3 times 1500 R.*

in the left leg with 62 500 Landschutz ascites cells and later irradiated with a single dose of 6000 R on this leg and with 2500 R on a 2 cm² area of the inferior left part of the abdomen. Before X-ray exposure, half of the mice were treated with a mixture of 5 HT + AET + cysteine and glutathione. This mixture of radioprotectors does not have any influence on the tumour growth in the non-irradiated protected mice. However, the survival of the irradiated protected mice was greater than that of the control and of the non-protected mice (Figs 7a and 7b). Indeed, within 100 days after the treatment, 95% of the irradiated protected mice and 92% of the normal control mice died, whereas 95% of the mice protected before irradiation were still alive. The causes of death in the first two groups of mice are not the same. All control mice present a tumour and die from generalized metastases. On the other hand, most of the irradiated non-protected mice died from X-ray lesions, presenting marked lesions in the small intestine and the bone marrow.

Moreover, it was seen that:

1. In the two irradiated groups, the percentage of tumours was very low;
2. In the irradiated non-protected mice the irradiated leg was necrotic, whereas in 40% of the protected irradiated mice the abdominal region and the right leg were normal.

Ascites cells grafted in the peritoneal cavity

Mice were injected in the peritoneal cavity with 31 500 ascites cells and irradiated on the abdomen with 3000 R divided into 4 doses of 1000, 900, 800 and 300 R given at 7-day intervals. The first irradiation dose was administered immediately after the graft of the tumour cells; the second and the fourth irradiations were followed by an injection of 10^7 of isologous bone-marrow cells. The protected mice were given 10, 8, 6 and 5 mg of AET-Br by stomach intubation before the 1st, 2nd, 3rd and 4th irradiation doses respectively.

The results on Fig.8 show that all irradiated non-protected mice die within 30 days and that 95% of the irradiated non-protected mice die within 55 days, whereas 55% of the protected mice were still alive at this time.

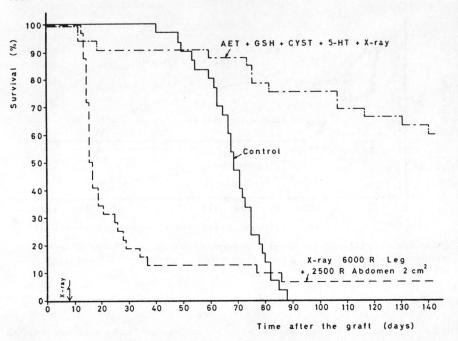

FIG. 7a. Survival of BALB/c mice grafted with 62 500 Landschutz ascites tumour cells in the left leg and irradiated on the same leg with a single dose of 6000 R and on 2 cm² of the inferior left part of the abdomen with a dose of 2500 R.

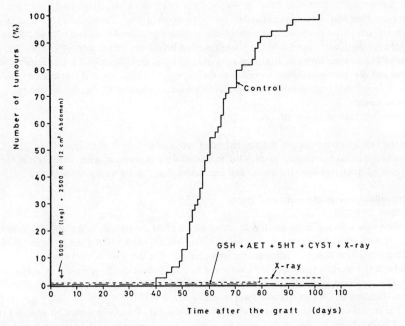

FIG. 7b. Percentage of cancers in control mice and in mice irradiated with or without protection (the experimental conditions were the same as in Fig. 8a).

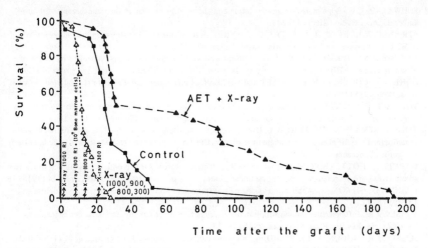

FIG.8. Survival of BALB/c mice grafted in the peritoneal cavity with 31 500 ascites cells and irradiated on the abdomen with 3000 R divided into 4 doses of 1000, 900, 800 and 300 R given at 7-day intervals.

DISCUSSION

The usefulness of sulphydryl radioprotectors in the radiotherapy of cancer is limited since (a) the range between the active and the toxic dose is narrow and (b) radioprotective drugs protect cancer tissues as well as normal ones. In a previous study, we have demonstrated that when mixtures of radioprotectors are given the degree of protection after 30 days is increased compared with that of AET alone and the toxicity is smaller [15, 20].

In addition, radioprotectors could be of use in the treatment of poorly vascularized tumours such as some tumours of the skin, the lymph nodes and the ovaries since the concentration of the sulphydryl radioprotectors in a given tissue depends essentially on its vascularization.

Our experimental results also suggest that radioprotectors may be useful in the radiotherapy of cancer when treatment of large parts of the body with medium doses combined eventuelly with high local doses is required to cure the tumour or to limit its spreading. However, it should be pointed out that since Landschutz tumours represent homografts, the cure achieved by X-rays may be due to cell killing as well as to an action on immunological factors leading to the rejection of the graft.

Before radioprotectors can be used on a large scale in the radiotherapic treatment of human cancer, it will be necessary to decrease still more their toxicity and to prevent the radioprotectors from reaching the cancer tissues. This might be realized in part by diminishing the dose of radioprotectors or by either reducing the time between the administration of the radioprotectors and the radiation exposure or administrating the radioprotectors in association with an inert compound such as thiogel a few minutes before irradiation. Another possibility would be to combine the treatment with the radioprotectors with that of a radiosensitizing compound. In this case, the radiosensitizing compound must be administrated in such a way that it reaches the tumour directly (for example by intra-arterial perfusion).

REFERENCES

[1] HALL, V.B., Preliminary Report: Cysteine protection of tumour fragments "in vitro" from irradiation with lethal dosages of X-rays (ANL-4401. Argonne Natl. Labor. Quarterly Rpt. Nov., Dec., 1949 and Jan. 1950). Biol. and Medical Diseases, 130.

[2] SCHWARTZ, E.E., Bone marrow transplantation and chemical protection in the radiotherapy of mouse leukemia, Acta Radiol. **50** (1959) 235.
[3] STRAUBE, R.L., PATT, H.M., SMITH, D.E., TYREE, E.B., Influence of cysteine on the radiosensitivity of Walker rat carcinoma-256, Prov. Am. Assoc. Cancer Res. **10** (1950) 243.
[4] ALTENBRUNN, H.J., HUBER, R., "Radiation susceptibility of animal tumours controlled by means of SH-compounds", VIIIth Int. Cancer Congress, Moscow, 1962, Acta Unio Int. contra Cancerum **19** (1962).
[5] COHEN, A., COHEN, L., Effects of AET and 5-HT on C_3H mammary tumour isografts irradiated "in vitro", Br. J. Radiol. **35** (1962) 200.
[6] TOLKACHEVA, E.N., Effects of radiations on Ehrlich ascites carcinoma in connexion with the problems of protection. I. Cellular character of the action of protective substance, Biophysics **4** (1959) 61.
[7] TOLKACHEVA, E.N., SHAPIRO, N.I., Use of protective substances as a method for differential changes in radiosensitivity of normal and malignant tissues, VIIIth Int. Cancer Congress, Moscow, 1962, Acta Unio Int. contra Cancerum **19** (1962).
[8] DARCIS, L., HOTTERBOX, S., De l'action radioprotectrice exercée sur la muqueuse rectale par la cystéamine administrée par voie générale et par la cystamine administrée par voie intrarectale Experientia **14** (1958) 18.
[9] DARCIS, L., GILSON, G., Application vaginale de cystamine et radio-protection locale, Experientia **13** (1957) 242.
[10] MAISIN, J.R., Influence des radiosensibilisateurs et des radioprotecteurs sur la réponse des cancers à la radiothérapie, Rev. Fr. Etud. Clin. Biol. **9** (1964) 437.
[11] MAISIN, J.R., HUGON, J., LEONARD, A., Radioprotecteurs et cancers, J. Belge Radiol. **47** (1964) 871.
[12] MAISIN, J.R., MATTELIN, G., Radioprotecteurs et radiothérapie des cancers, Bulletin du Cancer **54** (1967) 149.
[13] MAISIN, J.R., LEONARD, A., HUGON, J., Tissue and cellular distribution of tritium-labeled AET in mice, J. Natl. Cancer Inst. **35** (1965) 103.
[14] MAISIN, J.R., MATTELIN, G., LAMBIET-COLLIER, M., Reduction of short- and long-term radiation effects by mixtures of chemical protectors, Int. J. Radiat. Biol. (1970) 355.
[15] MAISIN, J.R., MATTELIN, G., FRIDMAN-MANDUZIO, A., VAN DER PARREN, J., Reduction of short- and long-term radiation lethality by mixtures of chemical protectors, Radiat. Res. **35** (1968) 26.
[16] MAISIN, J.R., Influence des radioprotecteurs sur le traitement radiologique des cancers, C.R. Séances Soc. Biol. **158** (1964) 193.
[17] MAISIN, J.R., LEONARD, A., Localisation de la 2-β-aminoéthylisothiourée dans les tissus cancéreux de la souris et influence de ce radioprotecteur sur le traitement radiologique des cancers, C.R. Séances Soc. Biol. **157** (1963) 671.
[18] MAISIN, J.R., LEONARD, A., Etude autoradiographique de la localisation de l'AET dans les tissus de la souris, C.R. Séances Soc. Biol. **157** (1963) 203.
[19] LEONARD, A., Analyse chromosomique d'une tumeur d'ascite de souris après 200 transplantations, C.R. Séances Soc. Biol. **158** (1964) 193.
[20] MAISIN, J.R., BACQ, Z.M., Toxicity, Ch. II of Sulfur-Containing Radioprotective Agents — International Encyclopedia of Pharmacology and Therapeutics (1975) 15.

NITROIMIDAZOLES AS HYPOXIC CELL SENSITIZERS IN VITRO AND IN VIVO

G.E. ADAMS, J.F. FOWLER
CRC Gray Laboratory,
Mount Vernon Hospital,
Northwood, Mddx.,
United Kingdom

Abstract

NITROIMIDAZOLES AS HYPOXIC CELL SENSITIZERS IN VITRO AND IN VIVO.
Of the many known radiosensitizers for hypoxic mammalian cells, the nitroimidazoles offer, at the present time, the best prospects for clinical use because of their favourable pharmacology. One of these, the 2-nitroimidazole, Ro-07-0582, is under preliminary clinical investigation. The current status of laboratory investigations with this drug, both in vitro and in vivo are reviewed. About twelve different in-vivo systems, including solid tumours in mice, have shown greater increased sensitivity to single doses of X-rays in the presence of this drug (dose-modifying factors of approximately 2). Several experimental tumours in mice have been tested with fractionated doses of radiation and drug and a therapeutic advantage was still observed. The 0-demethylated derivative of Ro-07-0582 is its major metabolite in all species investigated including man. In-vitro studies have shown that it is an efficient hypoxic cell sensitizer, although slightly less so than Ro-07-0582. It is, however, somewhat less toxic in mice. Several new nitroimidazoles have been tested for sensitizing ability in hypoxic mammalian cells cultured and irradiated in vitro. The results show that the 2-nitroimidazoles are more efficient than the 5-nitroimidazoles. The electron affinities of the sensitizers, expressed as one-electron reduction potentials, have been measured by pulse radiolysis and correlate well with the sensitization efficiencies of the compounds. The nitroimidazoles varied by a factor of 70 in their octanol/water partition ratio, but this variation has only a small effect on the sensitizing efficiency compared with the influence of electron affinity.

INTRODUCTION

The relative radioresistance of hypoxic cells is believed to be an important cause of local failure in the radiotherapy of malignant disease. Methods suggested to overcome the problem include treatment in hyperbaric oxygen chambers [1] and radiotherapy with heavy nuclear particles such as neutrons or negative π mesons [2]. A third approach is to use chemical agents which selectively increase the radiosensitivity of hypoxic cells without affecting the radiation sensitivity of well-oxygenated cells, i.e. most normal tissue [3]. The rationale of this approach is that these sensitizers are not rapidly metabolized and are able, therefore, to diffuse from the capillary vessels to distant hypoxic cells in tumours. There are many compounds which are active hypoxic cell radiosensitizers in vitro and it is established that the efficiencies of these compounds are a function of their electron affinities.

The search for compounds suitable for use as clinical sensitizers at first met with little success since most compounds investigated in vivo were either too toxic or were too rapidly metabolized. During this period, various drugs already in clinical use and possessing chemical structures associated with high electron affinities were investigated as potential sensitizers [4]. In 1973, it was reported that the trichomonacide, Metronidazole ("Flagyl"), was a hypoxic cell sensitizer [5–6]. Although in vitro this sensitizer is only moderately active on a concentration basis, experiments in various types of systems in vivo, including cure of solid tumours in mice [7], gave promising results. This was due to its low toxicity, wide distribution in tissues and particularly

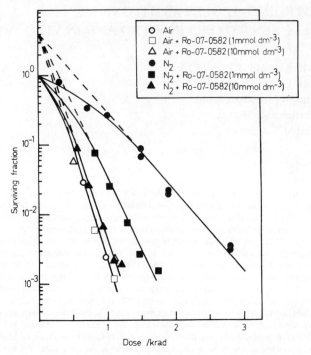

FIG.1. Survival data for oxic and hypoxic Chinese hamster V79-379A cells X-irradiated in the presence of 1 and 10 mmol · dm^{-3} Ro-07-0582 [11].

its long metabolic half-life. Preliminary clinical trials of Metronidazole as a radiosensitizer were, therefore, initiated [8—9]. At the same time, laboratory investigations were carried out with several related nitroimidazoles aimed at finding more active compounds [10].

The electron affinities of the 2-nitroimidazoles are higher than those of the 5-nitroimidazoles, e.g. Metronidazole, and one of these, the compound Ro-07-0582

$$\underset{\underset{NO_2}{|}}{\overset{N \diagdown \diagup N}{\bigvee}} - CH_2 \underset{|}{CH} \underset{|}{CH_2} \qquad \underline{Ro\text{-}07\text{-}0582}$$
$$\qquad\qquad\qquad OH \ OCH_3$$

has been shown to be a more effective sensitizer than Metronidazole. The current status of research on this drug is summarized in the following sections.

Sensitization by Ro-07-0582 in vitro

Figure 1 illustrates the sensitization of hypoxic Chinese hamster V79-379A cells cultured and irradiated in vitro in the presence of 1 and 10 mM Ro-07-0582 [11]. These and other data

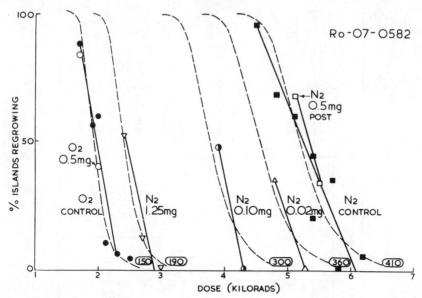

FIG.2. *Proportion of epidermal clones from basal cells surviving X-irradiation in vivo for various drug doses of Ro-07-0582* [16].

[11] show that the sensitization is only evident for hypoxic cells, is nearly equivalent to the full oxygen enhancement ratio for this cell line, 2.8, is independent of the serum concentration in the medium and is not affected by the position of the cell in the mitotic cycle.

Other investigations have shown that Ro-07-0582 [12–13] and Metronidazole [14] are much more toxic to hypoxic cells than they are to fully oxygenated cells. This specific toxicity, particularly when the drugs are in contact with the cells for several hours, is detectable in animal tumour experiments [15] and could make a useful contribution to the therapy of human tumours. However, so far, animal experiments show that this effect is smaller than the true radiosensitization afforded by these drugs.

Sensitization by Ro-07-0582 in vivo

(a) Artificially hypoxic mouse skin

Within the last two years, about sixteen in-vivo test systems in mice have been used to investigate radiosensitization of hypoxic cells by both Ro-07-0582 and Metronidazole. The first in-vivo test system used to demonstrate sensitization by these two drugs [16] was the skin clone method of assessing cell survival originally developed by Withers [17].

Mice were administered the drug at various concentrations before irradiation with an electron beam from a linear accelerator. Anaesthetized mice were rendered temporarily hypoxic by breathing nitrogen for about 35 seconds before and during irradiation. Test islands of skin were irradiated and the fraction of 'islands' which regrew as clones in situ were scored as a function of X-ray dose. The data are shown in Fig.2 for several doses of Ro-07-0582 expressed as mg/g body weight. Sensitization is shown by the progressive displacement of the percentage regrowth

TABLE I. RADIOSENSITIZATION OF MOUSE TUMOUR SYSTEMS BY RO-07-0582

Tumour	Doubling times (days)	Hypoxic fraction	Assay method	X-ray dose enhancement with Ro-07-0582		Author	
				(0.2–0.3 mg/g)	(1 mg/g)		
CBA fast sarc. F	1	10	Regrowth delay Loss of ^{125}IUdR	1.0	1.5	Begg	[19]
CBA ca. NT	3	6	Regrowth delay	1.4	2.2	Denekamp and Harris	[15]
WHT bone sarc. 2	2.5	–	Regrowth delay	–	1.8 [a]	Denekamp and Stewart	[20]
WHT Fibro-sarc.	2	–	Regrowth delay	–	1.8	Denekamp and Stewart	[20]
WHT sq. ca. D.	1	18	Cell dilution in vivo	(0.4 + 0.4) mg/g 1.0		Hewitt	[21]
CBA fast sarc. F	1	10	Cell dilution in vitro	1.3	2.2	McNally	[22]
WHT anap. MT line transplant	1	50	"	–	1.5	McNally	[22]
EMT 6	–	–	"	–	2.2	Har-Kedar et al.	[23]
EMT 6	–	–	"	–	2.4	Brown	[29]
C3H sarc. KHT	2	6	Cell dilution in vivo	1.2–1.3	1.8	Rauth	[24]

Tumour	Doubling times (days)	Hypoxic fraction	Assay method	X-ray dose enhancement with Ro-07-0582		Author	
				(0.2–0.3 mg/g)	(1 mg/g)		
WHT sq. ca. D	1	18	Cure (regrowth)	–	2.0[b] 2.0[b]	Hill and Fowler	[25]
WHT intradermal sq. ca. G	1	0.3	Cure	1.9	2.1	Peters	[26]
C3H 1st gen. trans. mamm. ca.	6	10	Cure	1.7	1.8	Sheldon et al.	[27]
WHT anap. MT line transplant	1	50	Cure	1.7	2.0[c]	Sheldon	[28]
C3H mamm. ca.	–	–	Cure	–	2.3	Brown	[29]
C3H 3rd gen. trans. of spont. mamm. ca.	–	–	Cure	–	2.4	Stone and Withers	[30]

[a] No post-effect
[b] Post-effect of 1.25–1.3
[c] Post-effect of 1.16

curves from the nitrogen control (no drug) towards the oxygen control curve (left). The solid lines are 'by eye' fits and the dashed lines are computer fits. Enhancement ratios of 2.2, 1.36 and 1.12 were obtained for respective drug doses of 1.25, 0.1 and 0.02 mg/g body weight. In similar experiments with Metronidazole, the enhancement ratios were 1.34 and ~ 1.25 for drug doses of 1.0 and 0.1 mg/g respectively. Figure 2 shows that no enhancement of cell killing occurred when the drug is given to oxygen-breathing mice, nor in nitrogen-breathing animals when the drug was given post-irradiation. Although no significant degree of sensitization by Ro-07-0582 has been observed in normal oxygenated skin either by the clonal regrowth method or by gross skin response [18], a small enhancement can be expected in normal tissues which contain some hypoxic cells, such as mature cartilage. This would have to be considered when planning radiotherapy with these drugs, as is also true for radiotherapy with either hyperbaric oxygen or with neutron beams.

(b) Solid tumours in mice

The mouse tumour systems which have been used to investigate sensitization by Ro-07-0582 and, in some instances, Metronidazole, differ widely in tumour type, growth rate and hypoxic cell fractions and involve several different methods of assessing the degree of sensitization. These include regrowth delay, loss of isotopically labelled ^{125}IUdR, cell-dilution assays and local control of tumours, i.e. 'cure'.

Table I shows the sensitization results from 16 different studies involving 12 different mouse tumour systems. Some of the data are taken from reports in the literature and the remainder are taken from studies currently in progress. All the studies except one show significant sensitization with Ro-07-0582. For drug doses of 1 mg/g body weight, most of the tumour systems give sensitization ratios (i.e. X-ray dose-modifying factors) of the order of 2. Even for the lower drug doses of 0.2 – 0.3 mg/g, ratios as high as 1.7 – 1.9 can be achieved. In two of the tumour systems where the sensitization was assessed by local control and in two others where regrowth delay was used, a significant effect of Ro-07-0582 was observed, even when the drug was given post-irradiation averaging about 1.2 DMF. There is evidence from in-vitro studies that the electron-affinic sensitizers are much more cytotoxic to hypoxic than to oxic cells if left in contact with them for several hours, and this desirably specific toxicity may be responsible for the small, but significant, sensitization evident when the drug is present only after irradiation.

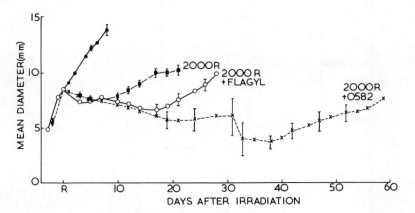

FIG.3. *Growth curves for the transplantable carcinoma NT in mice irradiated with 2000 rad of X-rays 15 min after the mice received 0.75 mg/g (i.p.) of Metronidazole and 2000 rad given 15 min after receiving 1 mg/g (i.p.) Ro-07-0582* [15].

Figure 3 shows some typical results for the effect of both Ro-07-0582 and Flagyl on radiation-induced regrowth delay in the mouse carcinoma 'NT' [15]. The growth delay caused by an X-ray dose of 200 rad was increased by about 12 days when 0.75 mg/g of Metronidazole was administered 15 min before irradiation. With a somewhat larger dose of Ro-07-0582, i.e. 1 mg/g, regrowth delay was increased to at least 35 days. The enhancement ratios given in Table I were obtained from a series of regrowth curves carried out for a range of X-ray doses. Some control experiments were carried out where the sensitizer was added **after** irradiation with 2000 rad. The additional delay was about 6 days, i.e. a significant effect, but smaller than the true sensitization evident when the drug is given before irradiation. As mentioned already, this post-effect is probably due to the cytotoxic effect of 0582 on hypoxic cells which predominate in the response of a tumour after a large single dose of X-rays.

Figure 4 shows 'cure' data for first-generation transplants of spontaneous mammary tumours in C3H/He mice irradiated with single doses of X-rays after administration of 1 mg/g body weight of Ro-07-0582 [27]. The tumours were irradiated at a mean diameter of 6.5 ± 1 mm and the drug was given intraperitoneally 30 min before irradiation. The ordinate is the proportion of tumours controlled, i.e. tumours which did not recur within 150 days. Although the X-ray dose enhancement ratio of 1.8 is fairly modest compared with some of the other ratios exceeding 2, the use of the drug increases the cure rate from 10% to about 90% for a radiation dose of 3200 rad. Further, the dose response curve for 0582 is significantly steeper than the control curve, almost in the same ratio of 1.8, suggesting strongly that, in these experiments, the drug reached all the hypoxic cells in the tumour and sensitized them efficiently. This X-ray enhancement ratio of 1.8 compares favourably with the gain factors of about 1.7 which have been determined for neutron or negative pion radiation.

Figure 5 shows similar data for local control of an anaplastic transplanted tumour 'MT' in WHT/Ht mice [28]. 'Cure' curves were obtained for drug doses of 0.1, 0.3 and 1.0 mg/g body weight. The enhancement ratio for the largest drug dose was 2.0, which is one of the largest

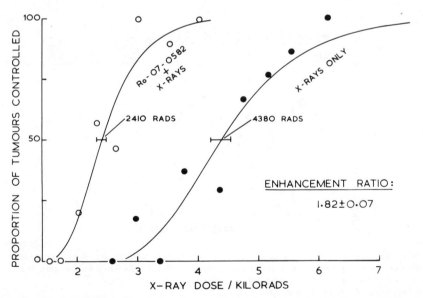

FIG.4. Proportion of C3H mice with transplanted mammary tumours cured as a fraction of dose. Right-hand curve — X-rays only; left-hand curve — X-rays starting 30 min after i.p. injection of 1 mg/g body weight of Ro-07-0582 [27].

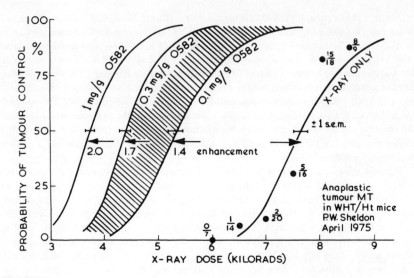

FIG. 5. *Proportion of WHT mice with transplanted tumours (MT) cured at 60 days as a fraction of radiation dose. Right-hand curve – X-rays only. The other three curves are for three different dosages of Ro-07-0582 given i.p. 30 min before starting irradiation [28].*

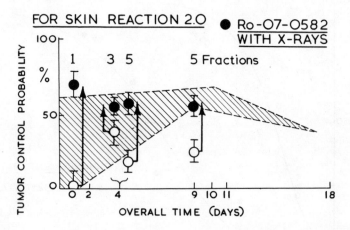

FIG. 6. *Variation of the proportion cured of C3H mice with transplanted mammary tumours treated with different schedules of fractionated X-rays with and without Ro-07–0582 (0.67 mg/g body weight). In each case the dose was that which produced a skin reaction of 2.0.*
Φ *X-ray alone,* ● *X-ray with Ro-07-0582*
The vertical arrows indicate the improvement in tumour control with respect to the concurrent X-ray-only experiment. The shaded area represents the envelope of results for other fractionated X-ray-only schedules using 2, 5, 9 or 15 fractions [31].

enhancements achieved by any agent in solid experimental tumours. Ratios of 2.3 – 2.4 have been reported from other laboratories (Stone and Withers; M. Brown). However, a drug dose of 1 mg/g is greater than would be suitable for use in clinical radiotherapy. The more practical concentration range is indicated in Fig.5 by the shaded area, which shows the range of enhancement ratios in mice for drug levels in serum equal to those that can be achieved in man, for single dose treatments. Even for the lowest enhancement ratio of 1.4, achievable with drug doses of 0.1 mg/g body weight, the probability of tumour control was increased from 5% to 95% in this tumour. In summary, the consistent sensitization shown in the data in Table I for a large number of mouse tumours implies that hypoxic cells are common to most mouse tumours and are the limiting factor in determining the curative dose levels. The presence of hypoxic cells in tumours is due to the biochemistry of oxygen metabolism rather than to any specific differences in tumour type. For this reason alone, the generality of the sensitization in mouse tumours is indicative of the extent to which hypoxic cells are likely to occur in human tumours.

Fractionated X-ray with Ro-07-0582

The large enhancement ratios obtained with hypoxic cell sensitizers and large single doses of X-rays will naturally be reduced in multiple-radiation dose treatments. This is mainly because reoxygenation of hypoxic cells can occur between radiation fractions and, if the fractionation schedule is such that reoxygenation is optimal, the hypoxic cells may be eliminated by the radiation alone without the necessity of sensitizers.

The C3H mouse mammary tumour system used in the study of sensitization by Ro-07-0582 has also been used for extensive investigations of optimum fractionation conditions for both X-rays [31] and neutrons [32]. Figure 6 reproduces some of the local control data for 3 and 5 fractions of X-rays given over 4 days and 5 fractions given over 9 days. In each case, the radiation doses were such that a constant skin reaction (averaging 2 on a scale of 3) was produced. Thus, the higher the experimental point on the ordinate scale of Fig.6, the greater is the therapeutic ratio. In two of the three schedules, 5 fractions in 4 or 9 days, the results for X-rays alone are mediocre and are the result of inadequate reoxygenation of hypoxic cells between fractions.

The vertical curves in Fig.6 show the improvement in local control when 0.67 mg/g body weight of Ro-07-0582 is given 30 min before starting irradiation. It is impressive that all four schedules, including the large single dose of X-rays, all bring the tumour control up to about the same level of 55 – 60% tumour cure. These results suggest that electron-affinic sensitizers, like neutrons, take the variability out of fractionated X-ray schedules. If the fractionation schedule is such that reoxygenation is not optimal, the remaining hypoxic cells are sensitized by the drug. If, however, the reoxygenation is efficient, sensitization becomes less important. Sensitizers would thus be particularly useful for non-standard X-ray schedules in clinical radiotherapy employing fewer and larger fractions because these appear to give poorer and more variable results than those obtained with regimes using many small fractions of X-rays alone.

Sensitizers with neutron irradiation

In a recent study, Denekamp et al. [32] compared the relative efficiencies of fractionated X-rays alone, fractionated X-rays with Ro-07-0582 and neutron irradiation on a mouse carcinoma 'NT' using regrowth delay as an end-point. As was found with the C3H mammary carcinoma, 0582 gave a large enhancement ratio of 2.1 with single doses of X-rays which, with fractionated doses, was reduced to 1.6 for 2 fractions in 48 h and to 1.3 for 5 fractions in 9 days. The reduction in enhancement ratio was due mostly to reoxygenation and partly to the reduced drug doses tolerated in multiple treatments (0.67 mg/g). These enhancement ratios were very close to the gains obtained in similar experiments using fast neutrons alone instead of X-rays.

The therapeutic gain obtained with fast neutrons is due mainly to the reduction in the size of the oxygen effect with this type of radiation. However, this adverse oxygen effect is not entirely eliminated and, therefore, it would be anticipated that if a hypoxic cell sensitizer were present during neutron irradiation, an even greater benefit would be observed. Various in-vitro studies with sensitizers using both bacterial and mammalian cell systems have demonstrated the additive effect of neutrons and sensitizers.

Fowler et al. [31] have also demonstrated additivity of 0582 and single doses of neutrons using the mouse tumour carcinoma NT and the regrowth delay method of assay. Significantly, no fractionated schedule either with neutrons or X-rays plus 0582 was better than the single neutron dose with the sensitizer present. Experiments with fractionated neutron doses plus 0582 are currently being done.

Optimum timing of drug and radiation treatment

Factors other than the inherent efficiency and dose-level of a given drug can influence the size of the therapeutic gain obtained with sensitizers. There is a large variation in the stability of sensitizers in vivo, some are metabolized extremely rapidly and some, e.g. Metronidazole and Ro-07-0582, are metabolized relatively slowly. Clearly, the longer the clearance time of the drug, the greater is the likelihood of the concentration of the drug in the hypoxic tumour cells building up to the level in the serum. There is little information available on either the rates at which sensitizers diffuse from blood capillaries in tumours to the distant hypoxic cells or on the various factors which might affect such diffusion rates. Lipophilicity and protein-binding equilibria may be important in this respect.

In determining, therefore, optimum timing between administration of sensitizer and irradiation of the tumour, it will be necessary to accumulate information from a wide range of animal tumours. An added complication lies in the large differences between the clearance half-lives in mammals of different species. The half-life of Ro-07-0582, for example, extends from about $1 - 1\frac{1}{2}$ h in mice through 4 h in the dog to $10 - 18$ h in man.

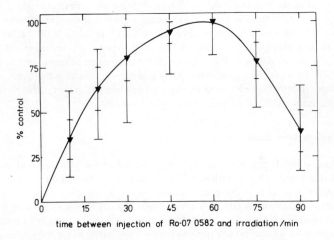

FIG. 7. *Proportion of tumours (MT) locally controlled at 80 days after 5000 rad of X-rays given at various times after i.p. injection of 0.2 mg/g Ro-07-0582 [28].*

Just how critical the timing might be is evident in some recent results of Sheldon [28] reproduced in Fig. 7. Batches of mice with the WHT anaplastic MT tumour were irradiated with single test doses of X-rays (5000 rad) at various times after administration of 0.2 mg/g body weight Ro-07-0582 and subsequently scored for local control. Figure 7 shows the local control probability for a constant dose of 5000 rad as a function of the time between dosage with the drug (i.p.) and the start of the irradiation. In the diagram, the inner error bars are standard derivations and the outer bars are the 95% confidence limit.

For this tumour system, the effectiveness of Ro-07-0582 as a sensitizer appears to be at a maximum when the drug is administerd 1 h before the irradiation. The increase in the proportion of tumours controlled during the first hour is most likely due to the build-up of the drug in the hypoxic tumour cells. The marked fall-off of tumour control during the next 30 min reflects the decrease, due to metabolism, of the 'pool' of Ro-07-0582 in the serum. By contrast, Denekamp [33] observed little or no variation in radiosensitization with different times between drug administration and irradiation for two different tumours, CBA carcinoma NT and WHT fibrosarcoma. In the first tumour the sensitizing ratios achieved with drug doses of 1.0 mg/g and 0.1 mg/g given 15 min before the start of irradiation were the same as those found for 1-h delay. Similarly, in the second tumour there was no difference between 15-min and 30-min delays for drug doses of 1.0 mg/g and 0.5 mg/g.

Clearly, much more work with different tumour systems is required to see how variable the optimum time really is. However, one can be fairly confident that the timing will not be so critical in the clinical situation, because of the much longer clearance half-life of the drug in man compared with the mouse. Evidence on this point is available in the results of recent pilot clinical studies with Ro-07-0582 carried out by Dische, Gray and Thomlinson [34–36]. This work is summarized in the paper by Lenox-Smith and Dische (these Proceedings).

DEVELOPMENT OF NEW HYPOXIC CELL SENSITIZERS

The general relationship between the electron affinities of the hypoxic cell sensitizers and their efficiencies has been established by various basic studies. These include pulse radiolysis investigations of one-electron transfer reactions between known sensitizers and the correlation of relative electron affinities of these compounds obtained from these studies with their efficiencies as hypoxic cell sensitizers for both prokaryotic and eukaryotic cells irradiated in vitro. One approach used in the nitrobenzene class of sensitizer has been the correlation of Hammett σ constants for the substituents in the benzene ring, with the sensitization efficiencies determined in Chinese hamster V79 cells in vitro [37]. However, a more direct approach would be to compare the sensitizing efficiencies with a more quantitative measure of the relative electron affinities. A trend has been observed [38] between sensitization efficiencies for several sensitizers of anoxic bacterial spores and reduction potentials taken from the literature, although it is now known that some of these potentials are not true one-electron potentials. However, accurate one-electron reduction potentials can now be determined by a pulse radiolysis method for measuring one-electron transfer equilibria between a solute and a reference compound of known reduction potential [39–42].

The one-electron potentials for a range of 2-nitro- and 5-nitroimidazoles have been measured by this method for comparison with the sensitizing efficiencies of the compounds determined for Chinese hamster cells irradiated in vitro [11]. Some of the results and conclusions from this study are discussed here.

For each compound, the sensitizing efficiencies were obtained from cell survival data, similar to those shown in Fig. 1, for a range of drug doses. The enhancement ratio can be calculated from the relative slopes of the linear portion of the survival curves for cells irradiated with and without sensitizer present. Figure 8 shows the enhancement ratios for four different 2-nitroimidazoles, which differ only in the nature of the N-1 substituted side chain, as a function

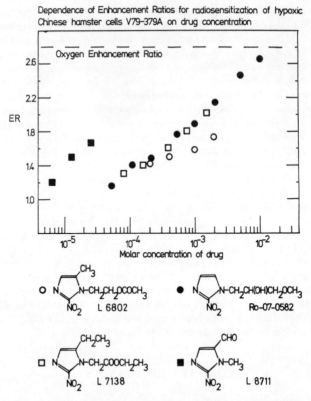

FIG.8. *Dependence of the enhancement ratios on drug concentration for sensitization of X-irradiated hypoxic Chinese hamster cells V79-379A by four different nitroimidazoles* [11].

TABLE II. STRUCTURES, ONE-ELECTRON REDOX POTENTIALS AND SENSITIZING EFFICIENCIES OF NITROIMIDAZOLES

Compound	R^1	R^2	E_7^1/mV [a]	$[S]$ [b] 1.6/mmol · dm^{-1}
L 8711	CH_3	CHO	− 243	0.02
L 7138	$CH_2CO_2CH_2CH_3$	CH_3CH_2	− 388	0.35
Ro-05-9963	$CH_2CH(OH)CH_2OH$	H	− 389	0.9
Ro-07-0582	$CH_2CH(OH)CH_2OCH_3$	H	− 389	0.3
Ro-07-0554	CH_2CH_2OH	H	− 398	0.3
L 6802	$CH_2CH_2OCOCH_3$	CH_3	− 420	1.0
L 6678	CH_2CH_2OH	CH_3	− 423	1.0
Metronidazole	CH_2CH_2OH	CH_3	− 486	4

[a] One-electron redox potential at pH7

[b] Concentration of sensitizer required to obtain an enhancement ratio of 1.6

R^1 and R^2 are the substituents at the N-1 and C-5 positions respectively

of drug concentration in the medium. Two of the compounds are ester derivatives, one of which is as efficient, on a concentration basis, as Ro-07-0582. The other appears to be somewhat less efficient. The most striking effect of the influence of the side chain on sensitizing properties is demonstrated by the 5-formyl derivative, L 8711, which is about 15 times more efficient than 0582. However, this drug is also very cytotoxic, which prevented the measurement of sensitization enhancement ratios for concentrations greater than about 25 μM.

In the study, several other nitroimidazoles were investigated and some of the results are summarized in Table II. The sensitizing efficiencies, defined as the concentration of drug in the medium required to produce an enhancement ratio of 1.6, were obtained from plots similar to those shown in Fig.8. Table II shows that the efficiencies cover a range of about 200 in concentration and it is noteworthy that all the 2-nitroimidazoles are more efficient than the 5-nitroimidazole, Metronidazole.

Also shown in the table are the one-electron reduction potentials measured by pulse radiolysis [42] using the method described by Meisel and Neta [41]. Observation of the one-electron transfer equilibrium

$$S^- + Q \rightleftharpoons S + Q^- \tag{1}$$

between the sensitizer radical-anion and a redox indicator such as quinone, enables the diffusion in redox potentials at pH7 Δ E, to be calculated for the two compounds. This difference is related to the equilibrium constant for the reaction, by the equation

$$\Delta E/mV = 59 \log K \tag{2}$$

Since the potential of the standard is known, that for the sensitizer can be calculated and, therefore, by selection of standards of appropriate reduction potential, data for a wide range of sensitizers can be obtained.

The results in Table II show that the radiosensitizing efficiencies correlate well with the measured one-electron reduction potentials. The octanol/water partition coefficients measured for these sensitizers, as an indication of their relative lipophilicities, varied over a range of about thirteen. This property does not seem to have a major influence on the sensitizing efficiency although there is evidence of some effect with the sensitizer Ro-05-9963. This compound is a urinary metabolite of Ro-07-0582 [44] and differs from 0582 only in that it is 0-demethylated. The reduction potentials of both drugs are identical, but the partition coefficient of 9963 is 0.11 compared with 0.43 for 0582. The table shows that there is a factor of three difference in sensitization efficiency.

Comparison of nitroimidazoles with other hypoxic cell sensitizers

The use of Hammett σ constants are useful for correlation of sensitization efficiencies. However, it is restricted to compounds with benzenoid structures. The availability of one-electron reduction potentials, however, enables correlations to be looked for in sensitizers of widely-differing chemical structure.

Figure 9 shows a plot of the sensitizing efficiencies of nitroimidazoles as a function of their one-electron reduction potential and includes data for oxygen, a nitrofuran, 5-nitro-2-furaldoxime [4] and the nitrobenzene, paranitroacetophenone (PNAP) (data from [43]). The sensitizing efficiencies of all three lie close to the line fitted to the results for the nitroimidazoles.

The implications of the correlation which holds over a range of sensitization efficiencies of about 2000 are considerable. Sensitizing efficiencies in vitro should be reasonably predictable provided that no other factors are involved in the sensitizing behaviour of a particular compound.

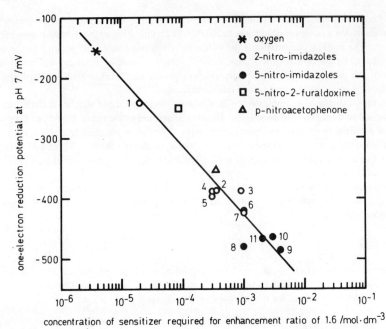

FIG.9. *Dependence of sensitization efficiency in V79-379A cells on electron affinity (one-electron reduction potential at pH7)* [11].
★ *Oxygen E_7^1 for standard state (1 mol · dm^{-3} oxygen).*

There are a few compounds that appear to be able to sensitize by combining oxygen-mimetic sensitization with effects unrelated to the oxygen effect. However, the vast majority of hypoxic cell sensitizers appear to be simple oxygen mimetics.

The octanol/water partition coefficients of the compounds represented in Fig.9 vary by a factor of over 300 strengthening the conclusions from the nitroimidazole data that lipophilicity is not an important factor in vitro. However, in vivo, it could play a much greater role, particularly in affecting the rate at which sensitizers diffuse from the tumour capillaries to the distant hypoxic cells as well as influencing the pharmaco-kinetics of the drugs. This is an area of important future investigation.

Finally, recent developments on the differential hypoxic-cell toxicity observed with a few sensitizers make it worth while to investigate a possible correlation between hypoxic cytotoxicity and electron affinity.

Much remains to be done, but there are good grounds for anticipating that clinical sensitizers even better than Metronidazole and Ro-07-0582 will become available.

ACKNOWLEDGEMENTS

It is our pleasure to thank our collaborators both in the Gray Laboratory and elsewhere for permission to quote extensively from their unpublished work.

REFERENCES

[1] The Biological Basis of Radiotherapy, Br. Med. Bull. **29** (1975) (various papers).
[2] FOWLER, J.F., in Proc. XIII Int. Cong. Radiology, Madrid, 1973, Excerpta Medica, Amsterdam (1973) 436.
[3] ADAMS, G.E., Br. Med. Bull. **29** (1973) 48.
[4] CHAPMAN, J.D., REUVERS, A.P., BORSA, J., PETKAU, A., McCALLA, D.R., Cancer Res. **32** (1972) 2616.
[5] FOSTER, J.L., WILLSON, R.L., Br. J. Radiol. **46** (1973) 234.
[6] CHAPMAN, J.D., REUVERS, A.P., BORSA, J., Br. J. Radiol. **46** (1973) 623.
[7] BEGG, A.C., SHELDON, P.W., FOSTER, J.L., Br. J. Radiol. **47** (1974) 399.
[8] DEUTSCH, G., FOSTER, J.L., McFADZEAN, J.A., PARNELL, M., Br. J. Cancer **31** (1975) 75.
[9] URTASUN, R.C., CHAPMAN, J.D., BAND, P., RABIN, H., FRYER, C., STURMWIND, J., Radiology (in press).
[10] ASQUITH, J.C., WATTS, M.E., PATEL, K., SMITHEN, C.E., ADAMS, G.E., Radiat. Res. **60** (1974) 108.
[11] ADAMS, G.E., FLOCKHART, I.R., SMITHEN, C.E., STRATFORD, I.J., WARDMAN, P., WATTS, M.E., Radiat. Res. (submitted for publication).
[12] HALL, E.J., ROIZIN-TOWIE, L., Radiology (in press).
[13] MOORE, B.A., PALCIC, B., SKARSGARD, L.D., Radiat. Res. (in press).
[14] MOHINDRA, J.K., RAUTH, A.M., Cancer Res. (in press).
[15] DENEKAMP, J., HARRIS, S.R., Radiat. Res. **61** (1975) 191.
[16] DENEKAMP, J., MICHAEL, B.D., HARRIS, S.R., Radiat. Res. **60** (1974) 119.
[17] WITHERS, H.R., Radiat. Res. **32** (1967) 227.
[18] FOSTER, J.L., Br. J. Cancer (in press).
[19] BEGG, A.C., Radiat. Res. (submitted for publication).
[20] DENEKAMP, J., STEWART, F. (private communication).
[21] HEWITT, H.B. (private communication).
[22] McNALLY, N., Br. J. Cancer (in press and private communication).
[23] HAR-KEDAR, I., WATSON, J.V., BLEEHEN, N.M., (personal communication).
[24] RAUTH, A.M., in Symp. on Chemical Radiosensitization of Hypoxic Cells (Proc. 5th Int. Cong. Radiat. Res. Seattle, 1974) (in press).
[25] HILL, S., FOWLER, J.F. (work in progress).
[26] PETERS, L. (private communication).
[27] SHELDON, P., FOSTER, J.L., FOWLER, J.F., Br. J. Cancer **30** (1974) 560.
[28] SHELDON, P (work in progress).
[29] BROWN, M., Radiat. Res. (in press).
[30] STONE, H.H.B., WITHERS, H.R., Br. J. Radiol. **48** (1975) 411.
[31] FOWLER, J.F., SHELDON, P.W., DENEKAMP, J., FIELD, S.B., Int. J. Radiat. Onc. Biol. Phys. (in press).
[32] DENEKAMP, J., HARRIS, S.R., MORRIS, C., FIELD, S.B., Radiat. Res. (submitted for publication).
[33] DENEKAMP, J. (private communication).
[34] GRAY, A.J., DISCHE, S., ADAMS, G.E., FLOCKHART, I.R., FOSTER, J.L., Clin. Radiol. (in press).
[35] DISCHE, S., GRAY, A.J., ZANELLI, G.D., Clin. Radiol. (in press).
[36] THOMLINSON, R.H., DISCHE, S., GRAY, A.J., ERRINGTON, L.M., Clin. Radiol. (in press).
[37] RALEIGH, J.D., CHAPMAN, J.D., BORSA, J., KREMERS, W., REUVERS, A.P., Int. J. Radiat. Biol. **23** (1973) 377.
[38] SIMIC, M., POWERS, E.L., Int. J. Radiat. Biol. **26** (1974) 87.
[39] ARAI, S., DORFMAN, L.M., in Adv. Chem. Ser. **82**, Am. Chem. Soc. (1968) 378.
[40] MEISEL, D., CZAPSKI, G., J. Phys. Chem. **79** (1975) 1503.
[41] MEISEL, D., NETA, P., J. Am. Chem. Soc. **97** (1975) 5198.
[42] WARDMAN, P., CLARKE, E.D., J. Chem. Soc. Farad. **1** (in press).
[43] ADAMS, G.E., ASQUITH, J.C., DEWEY, D.L., FOSTER, J.L., MICHAEL, B.D., WILLSON, R.L., Int. J. Radiat. Biol. **19** (1971) 575.
[44] FLOCKHART, I.R. (private communication).

N_2O-MEDIATED ENHANCEMENT OF RADIATION INJURY OF *Escherichia coli* K-12 MUTANTS IN PHOSPHATE-BUFFERED SALINE*

T. BRUSTAD, E. WOLD
Norsk Hydro's Institute for Cancer Research,
The Norwegian Radium Hospital,
Oslo, Norway

Abstract

N_2O-MEDIATED ENHANCEMENT OF RADIATION INJURY OF *Escherichia coli* K-12 MUTANTS IN PHOSPHATE-BUFFERED SALINE.
 The inactivation of radioresistant strains of *Escherichia coli* K-12 suspended in phosphate-buffered saline (PBS) is higher in suspensions irradiated under N_2O-bubbling than under N_2-bubbling. The sensitization is largely the result of OH-induced toxic products formed in N_2O-saturated PBS. Special emphasis is placed on establishing the condition under which extracellular toxic products with a life-time exceeding about 20 s are formed. It is concluded that chlorine atoms are the precursor for these toxic products and that N_2O and phosphate potentiate their formation.

INTRODUCTION

During the past years considerable effort has been put into attempts to modify the radio-response of cells. Recently, much work has been done to enhance specifically the radiation injury to anoxic and hypoxic cells as such cells are assumed, in certain cases, to limit the success of radiation therapy. Selective protection by certain chemicals of normal cells present within the radiation fields used to treat a tumour has also been attempted although less weight appears to have been given to this latter aspect during the last years. The effect of protective compounds is commonly interpreted as restitutive or competitive protection, whereas that of sensitizers is often explained according to the "complex-formation", "sulphydryl" or "recombination" hypothesis. Less emphasis has been put on the possibility that added compounds may affect the radiation chemistry whereby toxic products arise from otherwise inert compounds. As an example, the present paper shows that when common non-toxic constituents, like phosphate-buffered saline and nitrous oxide, are present during irradiation they give rise to products which are highly toxic, at least to bacteria.

MATERIAL AND METHODS

Phosphate buffers

The chemical composition of the phosphate-buffered saline (PBS) used in the present work was 1.78 g of $Na_2HPO_4 \cdot 2H_2O$, 1.36 g of KH_2PO_4 and 7.6 g of NaCl per litre of distilled water, giving a pH = 6.8 [1, 2]. In some of the experiments phosphate buffer without NaCl (PB) was used, otherwise the composition was the same as for PBS.

---

* This investigation was supported by grants from the Norwegian Cancer Society, the Norwegian Research Council for Science and the Humanities, and the Nansen Scientific Fund.

TABLE I. THE BACTERIAL STRAINS STUDIED

Bacterial strain E. coli K-12	Genotype	
	uvr	rec
AB 1157	+	+
AB 1886	A-6	+
AB 2463	+	A-13
JO-307	A-6	A-13

Bacterial strain

Four different strains of *Escherichia coli* K-12 were used; their genotypes are shown in Table I [2].

The single mutants AB 1886 and AB 2463 were selected because of their known deficiencies in excisional repair of DNA, controlled by the **uvr⁺** gene, and recombinational repair, controlled by the **rec⁺** gene. The strain JO-307 was selected because of its deficiencies in both excisional and recombinational repair. For repair capabilities, this strain becomes the counterpart of the repair-proficient wild-type strain, AB 1157.

Culture conditions

A single colony of the actual bacterial strain used was transferred to 5 ml of "YET-broth" medium and incubated over night at 37°C with aeration. The YET-broth contained per litre of double-distilled water 5 g of yeast extract, 10 g of tryptone, 10 g of NaCl and sufficient NaOH to attain pH = 7.0. Before irradiation the cells were washed twice in PBS and finally suspended in PBS (or, where specified, in PB). After irradiation the cells were again properly diluted in PBS and plated on "YET-agar" (5 g of yeast extract, 10 g of tryptone, 10 g of NaCl, 20 g of Bacto agar and sufficient NaOH to attain pH = 7.0). After incubating for about 20 hours at 37°C the fraction of bacteria surviving irradiation was determined from the number of visible colonies.

Irradiation

In some experiments the cell suspension was irradiated with 220 kV X-rays at room temperature at a dose-rate of approximately 2.2 krad/min, as determined by ferrous sulphate dosimetry [3]. The irradiation chamber employed was similar to that described by Howard-Flanders and Alper [4] It allows continuous bubbling of the cell suspension with any given gases before and during irradiation, and aliquots to be extracted without admitting atmospheric air or interrupting the gas flow. The gas passes through a water lock before leaving the exposure chamber to avoid back-diffusion of atmospheric air into the cell suspension. Specially purified gases were obtained from Norsk Hydro A/S. The oxygen content of the N_2 and N_2O gases used was less than 3 ppm and 50 ppm, respectively.

In some other experiments a 4-MeV electron linear accelerator was used as the radiation source In these cases dosimetry was performed with modified Fricke's ferrous sulphate dosimetry solution [5]. Details of the experimental set-up for electron irradiation are given in the text.

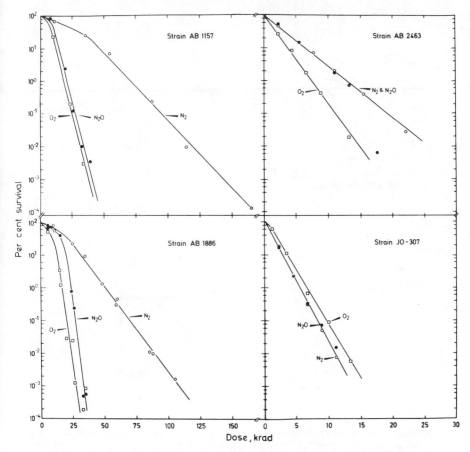

Fig.1. Dose survival curves of E. coli *K-12 strains AB 1157, AB 1886, AB 2463 and JO-307. The cells are irradiated suspended in PBS at concentrations less than 10^{7} ml^{-1}. The suspensions are bubbled with $\circ$ N_2, $\square$ O_2 and $\bullet$ N_2O before and during irradiation. Radiation: 220-keV X-rays.*

RESULTS

Figure 1 shows dose-survival curves of the four strains of *E. coli* K-12 after irradiation of cells suspended in PBS and bubbled with N_2, O_2 and N_2O. The data show that there is little or no N_2O-mediated enhancement of the lethality for the radiosensitive, recombination-deficient strains AB 2463 and JO-307. On the other hand, N_2O exerts a pronounced sensitization compared with the response in N_2 for the radioresistant, recombination-proficient strains AB 1157 and AB 1886. However, for doses of less than about 10 krad the enhanced radiation injury is rather small compared with that for higher doses. At first sight these results appear to be in contrast to earlier published results for vegetative cells. Thus, Müllenger and Ormerod [6] found N_2O to have no effect on the radioresponse of *Micrococcus sodonensis* and Ewing et al. [7], referring to unpublished data by Fielden, Lowlock and Brucks, state that "N_2O-sensitization has not been found with the several strains of vegetative cells tried". However, the composition of the phosphate buffer used by these authors is not specified.

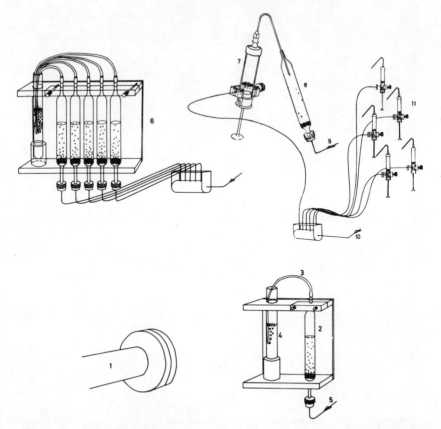

FIG. 2. Aliquots of the solution to be irradiated are contained in glass exposure vessels (2 and 6) where they are bubbled with the predetermined gas through the glass scinter filter. The gas is led in stainless-steel tubing from the high-pressure gas cylinders via reduction valves, flow-meters and O-ring sealed couplings to the glass exposure vessels. The effluent gas is led via water-locks (4). After irradiation with electrons from a linac (1), unirradiated cell suspension in glass syringes (11 and 8), prebubbled with a given gas, is manually injected into the irradiated solution in the glass exposure vessel (7, 8).

Further experiments showed that the N_2O-mediated enhancement of radiation injury of the resistant strains disappeared when 10^{-3}M of the effective OH scavenger t-butanol of $Fe(CN)_6^{4-}$ was added to the suspension. Thus, the effect appears somehow to be OH induced.

In an attempt to test whether long-lived toxic products are formed when N_2O-bubbled PBS is irradiated, the experimental set-up shown in Fig.2 was used. Aliquots of PBS were added to a number of irradiation glass vessels (2 and 6) where they were bubbled with N_2O (or O_2 or N_2) for 15 min. The effluent gas was led into a water-lock (3 and 4) to prevent back-diffusion of oxygen into the test solution. The PBS aliquots were subsequently irradiated with 4-MeV electrons from a linear accelerator (1) at a dose-rate of about 80 krad/s. Aliquots of bacterial suspension, contained in glass syringes with stainless-steel plungers (7 and 11) and shielded from irradiation, were bubbled with N_2 or air (10). At predetermined times after irradiation of a PBS sample the unirradiated cell suspension in a syringe was manually injected into the exposed buffer (7 and 8).

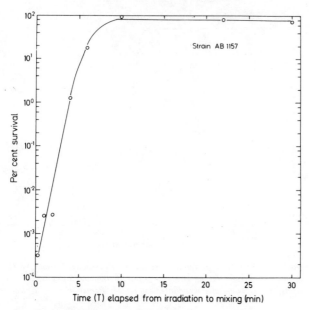

FIG.3. N_2O-saturated aliquots of PBS are irradiated (60 krad) and, at various times afterwards, admixed with N_2-bubbled (or air-bubbled) unirradiated aliquots of a suspension of E coli K-12, strain AB 1157. Ten minutes after mixing, the cell suspension was diluted and plated. Radiation: 4-MeV electrons.

Figure 3 shows results of experiments to determine the life-time of possible toxic products induced by irradiation of N_2O-saturated PBS. These experiments were carried out by irradiating (about 60 krad) N_2O-saturated aliquots of PBS and, at various times (T) afterwards, admixing N_2-bubbled (or aerobic) unirradiated aliquots of cell suspension AB 1157. Ten minutes after the admixing the samples were appropriately diluted in unirradiated PBS and plated.

These experiments show that the cell survival after a dose of about 60 krad is reduced by more than 5 orders of magnitude when T = 25 s, whereas it is reduced by less than 30% when T > 7 min. These results prove that toxic products are formed and that radiochemically they are to be considered as long-lived species with substantial killing ability at least for T < 7 min.

Experiments also showed that doses less than about 10 krad produced toxic products of so low concentration that hardly any reduction of cell viability in a suspension of 10^7 cells/ml was detectable.

Identical experiments were performed with the bacterial strains AB 1886, AB 2463 and JO-307. These experiments gave results similar to those presented here for the wild type strain AB 1157.

These results provide an explanation of the observation that little or no N_2O-mediated enhancement or injury occurred for the radiosensitive bacterial strains. The radiosensitivities of these strains are so high that even the maximum doses applied are too low to provide a yield of toxic products sufficiently high to reduce cell survival significantly. It also follows from these experiments that the enzymatic repair capabilities are not decisive for manifestation of damage caused by these long-lived toxic products.

Figure 4 shows results of experiments to test whether the toxic effect depends on the salt concentration in the phosphate buffer. In these experiments unirradiated bacteria are mixed with irradiated N_2O-bubbled PB supplemented with NaCl of various concentrations. The data show a pronounced toxic effect provided that the NaCl concentration exceeds 7.5×10^{-2} M, but little or no toxicity for lower cell concentrations.

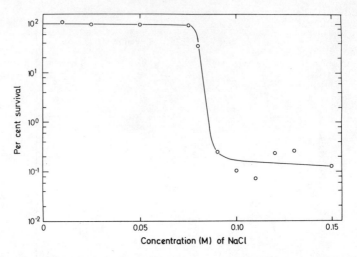

FIG.4. Per cent survival of unirradiated N_2-saturated suspensions of E. coli K-12, strain AB 1157 admixed with irradiated N_2O-saturated PB containing NaCl of various concentrations, about 20 s after irradiation of the latter. Ten minutes after mixing, the cell suspension was diluted and plated. Radiation: 4-MeV electrons.

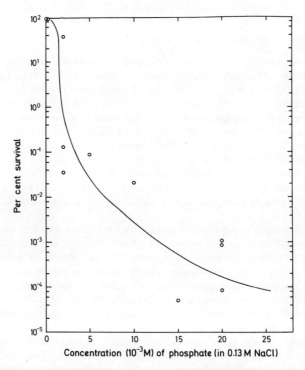

FIG.5. Per cent survival of unirradiated N_2-saturated suspensions of E. coli K-12, strain AB 1157 admixed with N_2O-saturated 0.13M NaCl solution containing various concentrations of phosphate, about 20 s irradiation of the latter. Ten minutes after mixing, the cell suspension was diluted and plated. Radiation: 4-MeV electrons.

Figure 5 shows results of experiments to test whether the toxic effect depends on the phosphate concentration added to an NaCl solution with the same NaCl concentration as PBS, i.e. 0.13M. In these experiments unirradiated bacteria were mixed with irradiated N_2O-bubbled 0.13M NaCl solution, supplemented with phosphate of different concentrations, but with the pH kept at 6.8. The data presented show that toxicity occurs provided the phosphate concentration exceeds about 2×10^{-3} M.

Experiments showed that there is no effect of long-lived toxic products when unirradiated bacteria, suspended in oxygenated or anoxic PB, are mixed with irradiated:

(1) N_2-bubbled PBS
(2) O_2-bubbled PBS
(3) N_2O-bubbled 2×10^{-2} M sodium acetate solution (with pH adjusted to 6.8 by CH_3COOH), irrespective of whether it contains 0.13M NaCl.
(4) N_2O-bubbled 1.3×10^{-2} M tri-sodium citrate solution (with pH adjusted to 6.8 by $C_6H_8O_7 \cdot H_2O$) supplemented with 0.13M NaCl.

The data presented above show that phosphate as well as NaCl are required in the N_2O-saturated solution to give rise to toxic products of sufficient concentration to affect cell survival.

Proposed mechanisms

Based on pulse radiolysis studies of aqueous sodium chloride solutions, Anbar and Thomas [8] and Ward and Myers [9] have proposed that a pH-sensitive conversion of OH radicals into chloride atoms occurs in a reaction which formally can be written as follows:

$$Cl^- + OH + H_3O^+ \rightarrow Cl^{\cdot} + 2H_2O \tag{1}$$

Details of reaction (1) are, however, not yet understood. We propose that $Cl^{\cdot}$ is the precursor of the toxic product formed.

In the absence of other solutes the Cl atoms combine with Cl^- to form Cl_2^-, which has an absorption maximum at 340 nm, namely [8, 9]:

$$Cl^{\cdot} + Cl^- \rightarrow Cl_2^- \tag{2}$$

By pulse radiolysis we have shown that transients having absorption maxima at 340 nm, and thus assumed to be Cl_2^-, are formed in N_2- as well as in N_2O-saturated PBS. These transients were found to decay similarly within a fraction of a second in N_2- and N_2O-saturated PBS. Since we have shown that the products giving rise to the toxic effect persist for minutes after irradiation of N_2O-saturated PBS, while they are absent in N_2-saturated PBS, it can be concluded that Cl_2^- is not directly responsible for the observed toxicity.

In analogy with known reactions for I^-, Br^- and $(CNS)_2^-$, it may be assumed that the decay of Cl_2^- is, in part, due to

$$Cl_2^- + Cl_2^- \rightarrow Cl_2 + 2Cl^- \tag{3}$$

In aqueous solution, Cl_2 is in equilibrium with hypochlorous acid through the reaction

$$Cl_2 + H_2O \rightleftharpoons HCl + HClO \tag{4}$$

The species Cl_2 and HClO are both known as highly active germicidal agents [10].

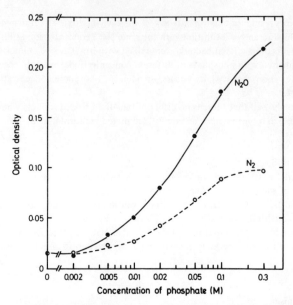

Fig.6. Relative yields of Cl_2^- ($\lambda = 350$ nm) in N_2O-saturated (●) and N_2-saturated (○) 0.13M NaCl solution supplemented with phosphate of different concentrations. All solutions were made up to pH = 6.8. Results are based on pulse radiolysis data 5 μs after irradiation. A 4-MeV electron accelerator was used. Pulse length 2 μs. Dose per pulse ~5 krad.

The scheme outlined through Eqs (1) to (4), however, does not explain the experimental fact that phosphate as well as NaCl is required in the N_2O-saturated solution to give rise to long-lived toxic transients of sufficient concentration to affect cell survival.

Our pulse radiolysis results, presented in Fig.6, show that the yield of Cl_2^- in 0.13M NaCl (pH adjusted to 6.8) is very low, and that there is no significant effect of N_2O on the yield. When phosphate buffer is added, however, the absorption at 350 nm increases with the phosphate concentration, and is more pronounced in N_2O- than in N_2-saturated solutions. Measurements of the absorption at 500 nm reveal that a negligible part of the increased absorption at 350 nm is caused by phosphate radicals, as explained further on.

The yield of Cl_2^- in N_2O-saturated PBS was found to be about five times as high as that in a neutral N_2O-saturated solution of 0.13M NaCl. These results demonstrate that the presence of phosphate somehow potentiates the formation Cl_2^-. Hence, according to Eqs (3) and (4), higher yields of the powerful germicides Cl_2 and HClO are expected in PBS than those in a neutral 0.13M NaCl solution. According to this expectation, Figs 5 and 6 demonstrate that the effect of phosphate on (a) cell survival and (b) on Cl_2^- yield becomes apparent at or near the same buffer concentrations.

Phosphate is generally considered to be radiochemically inert, although Pollard et al. [11] have proposed that radiation-induced short-lived phosphate radicals may potentiate the formation of strand breaks in DNA and RNA. Recent work of Grabner et al. [12, 13] has shown that both H- and OH-induced hydrogen abstraction from phosphate occurs. Thus, the following reaction

$$OH + H_2PO_4^- \xrightarrow{k_5} \cdot HPO_4^- + H_2O \qquad (5)$$

proceeds with a rate constant $k_5 = (2.0 \pm 0.5) \times 10^6 \cdot M^{-1} \cdot s^{-1}$ [13].

Based on recent pulse radiolysis studies of peroxodisulphate solutions by Redpath and Willson [14] we suggest, by analogy, that phosphate radicals from reactions like (5) participate in electron transfer reactions from chloride ions, such as

$$\dot{H}PO_4^- + Cl^- \rightarrow HPO_4^{2-} + \dot{C}l \qquad (6)$$

We have been unable to demonstrate unequivocally by pulse radiolysis studies that decay of phosphate radicals gives rise to a concomitant increase in the yield of Cl_2^- which, in subsequent reactions (Eqs (3) and (4)), might enhance the yield of Cl_2 and HClO. We propose, therefore, that a direct N_2O-mediated oxidation of $\dot{C}l$ occurs, possibly:

$$\dot{C}l + N_2O + H_2O \rightarrow ClOH + N_2 + \dot{O}H \qquad (7)$$

Since numerous reactions between oxidation products of chloride are known, the formation of species like Cl_2 and HClO may result in a variety of unstable, strongly oxidative and inter-related products. These reactive species will, in time, be converted into less reactive, stable oxidation products of chloride, a process which is supposed to explain the observed decay of toxicity with time elapsed after irradiation.

The lack of toxic products under nitrogen anoxia may, on the basis of the proposed mechanism, be explained partly by complete inhibition of reaction (7), and partly by back reactions of chlorine and phosphate radicals, like

$$\dot{C}l + e_{aq}^- \rightarrow Cl^- \qquad (8)$$

$$\dot{H}PO_4^- + e_{aq}^- \rightarrow H_2PO_4^- \qquad (9)$$

Oxygen, when present during irradiation, is expected to compete with essentially all the radical reactions outlined here. The lack of effect of long-lived toxic products after aerobic irradiation may, for instance, be explained partly by peroxidation of the phosphate radicals, a reaction which competes with and suppresses reaction (6), and partly by the complete inhibition of reaction (7).

The protective effect caused by OH scavengers present during irradiation may be explained as a suppression of reactions (1) and (5).

No effect of long-lived toxic products was observed when unirradiated bacteria suspended in PB supplemented with 10^{-3}M glutathione was admixed to irradiated N_2O-saturated PBS. This finding can be explained as reduction by glutathione of the highly oxidative toxic species, e.g. Cl_2 and HClO, hereby preventing them from reacting with the bacteria.

Sensitization by N_2O under conditions where long-lived toxic products are not formed

The present demonstration that long-lived toxic products are formed in irradiated N_2O-equilibrated PBS made it necessary to determine dose survival curves also under conditions where long-lived toxic products are not formed. The cells were, therefore, irradiated suspended in PB.

The data presented in Fig. 7 demonstrate that little (15% at the most) enhancement of radiosensitivity occurs from N_2O-mediated conversion of hydrated electron into OH radicals, under conditions where long-lived toxic products are not formed. The degree of N_2O-mediated sensitization is considerably lower than that reported by Powers and Cross [15] for bacterial spores. Thus, the N_2O-mediated sensitization observed when radioresistant bacteria are irradiated suspended in PBS is predominantly an effect of toxic products.

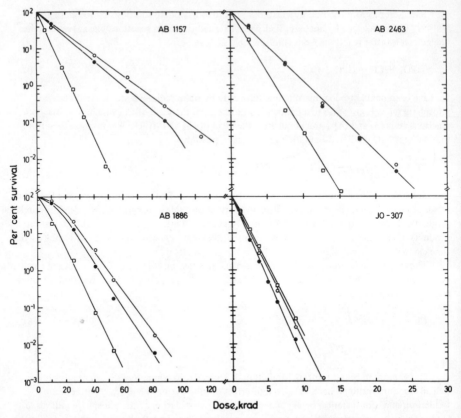

FIG.7. Dose survival curves of E. coli K-12, strains AB 1157, AB 1886, AB 2463 and JO-307. The cells are irradiated with X-rays while suspended in PB at concentrations of about $3 \times 10^6 \cdot ml^{-1}$. The suspensions are bubbled with ○ N_2, □ O_2 and ● N_2O before and during irradiation.

ACKNOWLEDGEMENTS

The technical assistance of Mrs. B. Mathiesen, Mrs. T. Naper and Mr. A. Rønnestad is gratefully acknowledged.

REFERENCES

[1] JOHANSEN, I., HOWARD-FLANDERS, P., Macromolecular repair and free radical scavenging in the protection of bacteria against X-rays, Radiat. Res. **24** (1965) 184.
[2] JOHANSEN, I., GULBRANDSEN, R., PETTERSEN, R., Effectiveness of oxygen in promoting X-ray induced strand breaks in circular phage λ DNA and killing of radiation-sensitive mutants of *Escherichia coli*, Radiat. Res. **58** (1974) 384.
[3] SHALEK, R.J., SINCLAIR, W.K., CALKINS, J.C., The relative biological effectiveness of 22 MeVP X-rays, Co-60 gamma rays and 220 kVP X-rays. II. The use of the ferrous sulphate dosimeter for X-ray and gamma-ray beams, Radiat. Res. **16** (1962) 344.
[4] HOWARD-FLANDERS, P., ALPER, T., The sensitivity of microorganisms to irradiation under controlled gas conditions, Radiat. Res. **7** (1957) 518.

[5] THOMAS, J.K., HART, E.J., The radiolysis of aqueous solutions at high intensities, Radiat. Res. **17** (1962) 408.
[6] MÜLLENGER, L., ORMEROD, M.G., The radiosensitization of *Micrococcus sodonensis* by N-ethyl maleimide, Int. J. Radiat. Biol. **15** (1969) 259.
[7] EWING, D., FIELDEN, E.M., ROBERTS, P.B., Modification of radiation sensitivity of *Bacillus megaterium* spores by N_2O and p-nitroacetophenone, Radiat. Res. **58** (1974) 481.
[8] ANBAR, E., THOMAS, J.K., Pulse radiolysis studies of aqueous sodium chloride solutions, J. Phys. Chem. **68** (1964) 3829.
[9] WARD, J.F., MEYERS, L.S., The effect of chloride ions on some radiation chemical reactions in aqueous solution, Radiat. Res. **26** (1965) 483.
[10] HADFIELD, W.A., "Chlorine and chlorine compounds", *in* Antiseptics, Disinfectants, Fungicides, and Chemical and Physical Sterilization"(REDDISH, G.F., Ed.), Lea & Febiger, Philadelphia (1957) 558.
[11] POLLARD, E.C., WELLER, P.K., Chain scission of ribonucleic acid and deoxyribonucleic acid by ionizing radiation and hydrogen peroxide *in vitro* and in *Escherichia coli* cells, Radiat. Res. **32** (1967) 417.
[12] GRABNER, G., GETOFF, N., SCHWÖRER, F., Pulsradiolyse von H_3PO_4, $H_2PO_4^-$, HPO_2^{2-} und $P_2O_7^{4-}$ in wässriger Lösung-II. Spektren und Kinetik der Zwischenprodukte, Int. J. Radiat. Phys. Chem. **5** (1973) 405.
[13] GRABNER, G., GETOFF, N., SCHWÖRER, F., Pulsradiolyse von H_3PO_4, $H_2PO_4^-$, HPO_4^{2-} und $P_2O_7^{4-}$ in wässriger Lösung-I. Geschwindigkeitskonstanten der Reaktionen mit den Primärprodukten der Wasserradiolyse, Int. J. Radiat. Phys. Chem. **5** (1973) 393.
[14] REDPATH, J.L., WILLSON, R.L., Chain reactions and radiosensitization: model enzyme studies, Int. J. Radiat. Biol. **27** (1975) 389.
[15] POWERS, E.L., CROSS, M., Nitrous oxide as a sensitizer of bacterial spores to X-rays, Int. J. Radiat. Biol. **17** (1970) 501.

RADIATION SENSITISATION BY MEMBRANE-SPECIFIC DRUGS*

M.A. SHENOY, K.C. GEORGE, V.T. SRINIVASAN,
B.B. SINGH, K. SUNDARAM
Bio-medical Group,
Bhabha Atomic Research Centre,
Trombay, Bombay, India

Abstract

RADIATION SENSITISATION BY MEMBRANE-SPECIFIC DRUGS.
 Recently, certain therapeutically active membrane-specific drugs have been shown to sensitise bacterial cells and mammalian cells in vitro to ionising radiations. Though sharing the property of membrane specificity, the radiochemical mechanisms of their action have been observed to vary. The synergistic sensitisation of bacterial cells by two of the compounds investigated has opened new vistas on the possibility of synergism between two or more radiosensitisers in clinical radiotherapy. The toxicology and pharmacology of these drugs are well documented and as such make experimental work on animal and human tumours easier. The experimental data on the use of some of these drugs on bacterial cells and mouse fibrosarcoma in vivo are presented. The possible applications of these studies in clinical radiotherapy are presented in relation to our observations.

1. INTRODUCTION

Studies with iodoacetic acid have shown that radiation sensitisation by iodine compounds is due to the liberation of iodine atoms ($I^{\bullet}$) during radiolysis of the sensitiser [1, 2]. It was further demonstrated that the radiolytically induced or chemically produced iodine atoms [3] combined with proteins in general and membrane proteins in particular, leading to the inhibition of post-irradiation repair [4], specifically the rejoining of DNA single-strand breaks [5].

The association between membrane damage and radiosensitisation prompted us to test certain therapeutically active drugs known to act on the cell membrane. As a result, a new group of radiosensitisers has evolved which includes anaesthetics, analgesics, hypnotics and tranquilisers. The advantage with this group of chemicals is their well-known toxicology and pharmacokinetics, thus facilitating the possible use of these chemicals in clinical trials.

Procaine hydrochloride, the common local anaesthetic, was the first drug in this series. This drug was selected because of its membrane specificity and also for the presence of a carbonyl group in its structure. On the basis of the electron affinity of the C = O group, Adams [6] has demonstrated radiosensitisation by many such compounds in a variety of test systems.

Apart from procaine hydrochloride, we have also screened many other drugs and present here data on the results of these studies in bacterial systems and mouse fibrosarcoma in vivo.

Recently, iothalamic acid has been demonstrated to sensitise single-cell systems to ionising radiations [7, 8]. This non-toxic organic iodine compound is frequently used for angiography and, as such, can be administered into human subjects in quite large doses without any ill effects. Here, we also present the results of our investigations with iothalamic acid as a radiosensitiser using mouse fibrosarcoma as the test system.

---

* This investigation was partly supported by the International Atomic Energy Agency, Vienna (Contract No. 1396/RB).

2. MATERIAL AND METHODS

2.1. Bacterial system

Escherichia coli B/r cells grown to stationary phase in nutrient broth (Difco) were suspended in 0.1M phosphate buffer (pH7) at a concentration of approximately 2×10^8 cells·ml^{-1}. The cell suspension was irradiated in a ^{60}Co gamma-ray source at a dose-rate of 6 krad·min^{-1} and the radiation dose was measured by ferrous sulphate dosimetry. The cell suspension was deoxygenated by bubbling oxygen-free nitrogen before and during irradiation. All the experiments were carried out at least twice with replicate plates at all plating dilutions. Irradiated samples were diluted in sterile phosphate buffer, inoculated into Petri plates and overlaid with nutrient agar (Difco) previously melted and held at 48°C. A minimum of 500 colonies of bacteria were counted at the end of 18 h incubation at 37°C. From the exponential part of the survival curves plotted on a semilog scale, D_1 (dose for 1% survival) values were obtained and the dose-modifying factor (DMF) was calculated as the ratio of D_1 in the presence of the chemical and that of control cells. Post-irradiation mixing was carried out within 20 s of irradiation.

Toxicity of the chemicals towards bacteria was determined up to a contact time of 100 min in buffer and only non-toxic concentrations were used in the present studies.

2.2. Mouse fibrosarcoma in vivo

A transplantable fibrosarcoma obtained from the Cancer Research Institute, Bombay was maintained in Swiss male mice (8 weeks old) by serial transplantation. The preparation of viable cells for transplantation was according to the method of Madden and Burk [9].

Two sets of experiments were carried out. In the first one, called single dose experiment, the individual radiosensitising effect of procaine hydrochloride and iothalamic acid was investigated. In the second set, hereafter referred to as double dose experiment, the radiosensitising effect of procaine hydrochloride on the tumours previously sensitised by iothalamic acid was studied.

2.2.1. Single dose experiments

For this experiment 2×10^6 viable tumour cells, as assessed by dye-exclusion test, were injected subcutaneously in the left hind limb of each animal. When the tumours reached a mean diameter of 1.05 ± 0.04 cm they were divided into five groups, each containing thirteen animals. Each animal in the first group was treated with 1 ml of procaine hydrochloride (0.25%) and the second with 1 ml of iothalamic acid (20%). The third and fourth groups received the same treatment and, in addition, were irradiated with X-rays. The fifth group treated with 1 ml of sterile distilled water and then irradiated served as the control. Irradiation was performed after five minutes of treatment. In all cases, the treatment was given by local injection at the tumour site. Local irradiation of the tumours was carried out using an X-ray machine operated at 250 kVp, 15 mA and using a 2-mm copper filter. The dose delivered was 2000 rad at a dose-rate of 200 rad·min^{-1}.

After irradiation, tumour sizes were measured regularly, the value of the square root of the product of length and breadth being taken as a measure of the size of the tumour.

2.2.2. Double dose experiments

For this experiment, 5×10^5 cells were transplanted in each animal. When the tumours reached a size of 0.78 ± 0.05 cm they were locally injected with 1 ml of 20% iothalamic acid and then irradiated to a dose of 3000 rad. When the tumour regressed to a minimum size, the animals

TABLE I. MODIFICATION OF RADIATION RESPONSE OF *E. coli* B/r BY SOME MEMBRANE-BINDING AGENTS

Gas	Sensitiser	Concn (mM)	D_1 dose (krad)	DMF
N_2	–	–	74.0	–
O_2	–	–	24.2	0.32
N_2	Lignocaine	10	49.0	0.66
O_2	Lignocaine	10	24.2	1.00
N_2	Tetracaine	1	51.0	0.69
O_2	Tetracaine	1	24.2	1.00
N_2	Meprobamate	25	65.0	0.88
O_2	Meprobamate	25	24.0	1.01
N_2	Sodium pentobarbitone	10	51.0	0.69
O_2	Sodium pentobarbitone	10	24.0	1.01
N_2	Chlorpromazine	0.1	40.5	0.54
O_2	Chlorpromazine	0.1	23.7	0.98

were divided into two groups. The animals in one group were injected with distilled water while those in the second received procaine hydrochloride. Both groups were then irradiated to a total dose of 3000 rad. The animals were monitored up to a period of 85 days for cure.

Tetracaine was bought from Glaxo Laboratories. Lignocaine was a gift from Suhrid Geigy, sodium pentobarbitone from Hoechst Pharmaceuticals, meprobamate from Geoffrey Manners and chlorpromazine from May and Baker, all of Bombay. All the chemicals were used as obtained from the manufacturers, without further purification.

Sodium nitrate and tertiary butanol were 'Analar' grade chemicals and were used at a concentration of 1mM.

3. RESULTS AND DISCUSSION

3.1. Bacterial system

E. coli B/r were sensitised to ^{60}Co gamma rays by lignocaine, tetracaine, meprobamate, sodium pentobarbitone and chlorpromazine under hypoxic conditions (Table I). As can be seen from the table, the drug concentration for the sensitising effect was fairly high except with chlorpromazine. Cells treated with chlorpromazine, but irradiated in the absence of the free chemical, showed a residual effect. It was also observed that adding irradiated chlorpromazine to unirradiated cells was ineffective, while mixing unirradiated chlorpromazine with irradiated cells showed a slight post-irradiation effect. Scavenging of hydrated electrons with sodium nitrate did not influence the sensitisation but scavenging of hydroxyl radicals with t-butanol resulted in reduced lethality. This effect was also observed in cells treated with chlorpromazine and then irradiated in the presence of t-butanol (Fig. 1).

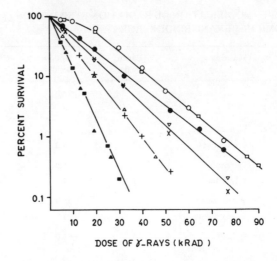

Fig.1. Radiation response of E. coli *B/r under different experimental conditions.*
□ —— □ *Anoxic buffer control*
■ —— ■ *Oxic buffer control*
○ —— ○ *Tertiary butanol control*
+ —— + *Cells + chlorpromazine − anoxic*
▲ —— ▲ *Cells + chlorpromazine − oxic*
▽ —— ▽ *Cells + chlorpromazine + t-butanol − anoxic*
× —— × *Cells treated with chlorpromazine − anoxic*
△ —— △ *Cells + chlorpromazine + NaNO$_3$ − anoxic*
● —— ● *Unirradiated chlorpromazine added to anoxically irradiated cells.*

The results obtained with chlorpromazine indicate that radiation sensitisation of *E. coli* B/r is due to some transients of the sensitiser formed by the reaction of hydroxyl radicals. The ineffectiveness of irradiated chlorpromazine when added to unirradiated cells and a very slight post-irradiation effect of chlorpromazine rules out any significant involvement of either the drug toxicity or inhibition of post-irradiation repair by chlorpromazine per se. Chlorpromazine has been shown to inhibit soluble and membrane-bound enzymes involved in respiratory chain and oxidative phosphorylation [10]. This may partly explain the sensitisation noticed in cells treated with chlorpromazine before irradiation. However, the presence of t-butanol during irradiation of chlorpromazine-treated cells reduced the residual effect, indicating the importance of hydroxyl radicals also, for residual sensitisation.

The fact that procaine can form salts or conjugate with other drugs and enhance the drug action has been known for a long time [11]. Consequently, it was of practical interest to see whether a combination of chlorpromazine and procaine would exert a synergistic sensitising action. When chlorpromazine (0.1mM) and procaine (25mM) were present during hypoxic irradiation of *E. coli* B/r in buffer, the sensitising effect obtained was identical to the effect produced by procaine alone (Fig. 2). Our earlier studies [12] have shown that about 80% of the total sensitisation by procaine could be obtained if it is added to cells immediately after irradiation. Therefore, when procaine (final concentration 25mM) was added to *E. coli* B/r irradiated under hypoxic conditions in the presence of chlorpromazine (0.1mM), the sensitisation was significantly greater than that produced even by oxygen. Conversely, when chlorpromazine was added to cells irradiated in the presence of procaine in anoxia, only the procaine effect was evident (Table II). As can be seen

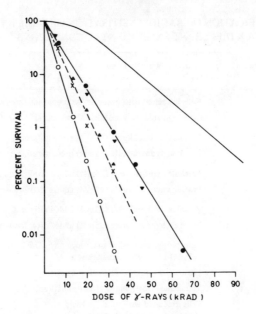

FIG.2. *Effect of procaine hydrochloride and chlorpromazine on* E. coli *B/r.*
———— *Anoxic buffer control*
- - - - - *Oxic buffer control*
● —— ● *Cells + procaine hydrochloride + chlorpromazine − anoxic*
○ —— ○ *Cells irradiated with chlorpromazine in anoxia; procaine hydrochloride added later*
× —— × *Cells irradiated with chlorpromazine in oxygen; procaine hydrochloride added later*
▲ —— ▲ *Cells irradiated in oxygen; procaine hydrochloride added later*
▼ —— ▼ *Cells irradiated with procaine hydrochloride in anoxia; chlorpromazine added later.*

from the results shown in the table, the post-irradiation effect of procaine was also observed when either lignocaine or tetracaine was substituted in place of procaine. The extent of post-irradiation sensitisation was identical with procaine concentrations of 2, 5, 10 and 25mM. The residual sensitisation caused by prior treatment of cells with chlorpromazine was also enhanced by the addition of procaine after irradiation. Addition of procaine to *E. coli* B/r cells irradiated in oxygen, or to cells irradiated with chlorpromazine in oxygen, demonstrated only the oxygen effect.

The absence of any additive effect when both procaine and chlorpromazine were present during irradiation can be explained in terms of scavenging of hydroxyl radicals by procaine which are responsible for the chlorpromazine effect. Although the relative rates of reaction of hydroxyl radicals with procaine and chlorpromazine are not available, the high concentration of procaine (25mM) with respect to chlorpromazine (0.1mM) would favour such an interpretation.

Since chlorpromazine by itself does not show any significant effect when added to irradiated cells, the ineffectiveness of adding chlorpromazine after irradiation to cells irradiated in the presence of procaine is self-explanatory.

But the post-irradiation treatment of cells with procaine irradiated in the presence of chlorpromazine could result in enhanced molecular deformation of the cell membrane and/or repair inhibition. We have earlier presented evidence [12, 13] that radiation sensitisation by procaine would be the result of both membrane damage and inhibition of post-irradiation repair. Alper [14] attributes cell death mainly to two types of primary events: type 'N' in nucleic acids and type 'O'

TABLE II. MODIFICATION OF RADIOSENSITISATION OF *E. coli* B/r BY CHLORPROMAZINE UNDER DIFFERENT EXPERIMENTAL CONDITIONS

Gas	Chlorpromazine (mM)	Experimental conditions	D_1 (krad)	DMF
N_2	–	Cells irradiated with procaine hydrochloride (25mM)	31.0	0.42
O_2	–	Cells irradiated with procaine hydrochloride (25mM)	23.5	0.98
N_2	0.1	Procaine hydrochloride (25mM) present during irradiation	31.5	0.42
O_2	0.1	Procaine hydrochloride (25mM) present during irradiation	23.5	0.98
N_2	0.1	Procaine hydrochloride (25mM) added after irradiation	15.1	0.20
O_2	0.1	Procaine hydrochloride (25mM) added after irradiation	23.2	0.98
N_2	–	Procaine hydrochloride (25mM) added after irradiation	38.5	0.52
O_2	–	Procaine hydrochloride (25mM) added after irradiation	24.3	1.00
N_2	0.1	Chlorpromazine washed off; procaine hydrochloride (25mM) added after irradiation	40.8	0.54
N_2	0.1	Lignocaine (10mM) added after irradiation	15.0	0.20
N_2	0.1	Tetracaine (1mM) added after irradiation	15.0	0.20
N_2	–	Cells irradiated with procaine HCl; chlorpromazine added after irradiation	31.8	0.42
O_2	–	Cells irradiated with procaine HCl; chlorpromazine added after irradiation	24.0	1.00
N_2	–	Cells irradiated with lignocaine; chlorpromazine added after irradiation	49.0	0.66
N_2	–	Cells irradiated with tetracaine; chlorpromazine added after irradiation	51.0	0.69

in membranous structures. Since DNA is attached to the membrane in *E. coli* the interplay of both 'N' and 'O' types of damage have a key role in the final expression of colony-forming ability of the bacterial cells. Both procaine and chlorpromazine are membrane-binding drugs. Consequently, one has to take into consideration the changes in molecular configuration and/or membrane distort brought about by these drugs. It is reasonable to assume at this juncture that such damage to membrane has to be repaired for the cells to survive. In fact, evidence for biochemical repair of membrane damage in *E. coli* B/r spheroplasts has already been presented [15]. Hence, the observat that procaine treatment even after irradiation is effective in enhancing the damage even beyond the oxygen effect implies that procaine alters the nature of damage to the cell membrane induced by chlorpromazine under hypoxia. The result of this would be either irrepairable damage or inhibition of specific repair processes.

Furthermore, damage to the membrane by iodine compounds [2] has also been shown to inhi post-irradiation DNA repair [4, 5] indicating that the structural integrity of the cell membrane is essential for the repair of DNA damage. Our own investigations [13] have ruled out any direct damage to DNA by procaine and a recent study [16] has shown increased damage to membrane and inhibition of the rejoining of DNA single-strand breaks, even when procaine is added after irradiatio

The absence of post-irradiation effect when procaine is added to aerobically irradiated cells either in the presence or absence of chlorpromazine implicates the involvement of differential damage and repair occurring when *E. coli* B/r cells are irradiated in the absence or presence of oxyg

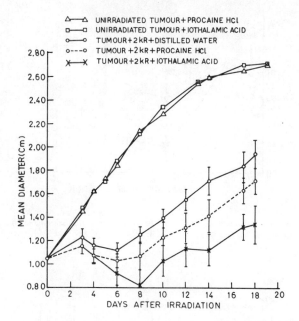

FIG.3. *Effect of procaine hydrochloride and iothalamic acid on post-irradiation growth of fibrosarcoma in mice. Dose — 2 krad.*

An alternative explanation would be a similarity in the action of oxygen and procaine, more so because procaine is ineffective in the presence of oxygen [12]. Lipid peroxidation is one of the mechanisms by which oxygen acts at the membrane level. Our own observations (P. Madon and B.B. Singh — unpublished data) also indicate that procaine reacts specifically with the lipid millieu of the membrane, as studied by NMR spectroscopy. Hence, if peroxidation by oxygen has saturated all the sites in the membrane where procaine too would act, then the subsequent addition of procaine after aerobic irradiation would be ineffectual in enhancing the damage further. We admit that this statement is rather conjunctural. Still, in view of our findings, this seems a most reasonable explanation for the absence of post-irradiation effect in the presence of oxygen.

The above observations indicate that the presence of chlorpromazine sensitizes *E. coli* B/r to ^{60}Co gamma rays via the hydroxyl-radical-induced transients and also by way of membrane damage. The subsequent addition of procaine after irradiation presumably fixes the membrane damage and also inhibits the rejoining of DNA single-strand breaks. The individual contributions of chlorpromazine and procaine, and also the mechanisms involved in these processes are currently under investigation.

3.2. Mouse fibrosarcoma in vivo

It can be seen from Fig. 3 that the unirradiated tumours grew progressively whereas the irradiated ones, after an initial growth, began to shrink and then regrow. The animals treated with procaine or iothalamic acid and then irradiated exhibited a reduction in the size of the tumours as compared with the controls, the effect being more pronounced with iothalamic acid. This can be attributed to the fact that it is known to sensitise both oxic and anoxic cells [7] while procaine

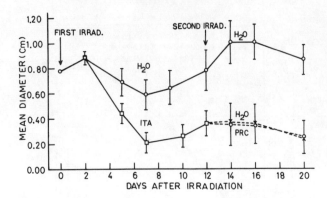

FIG.4. Effect of procaine hydrochloride on post-irradiation growth of fibrosarcoma in mice previously radio-sensitised by iothalamic acid. Dose — 3 krad for each irradiation.

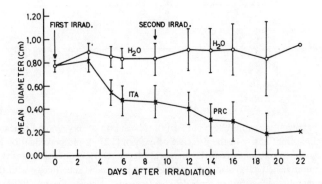

FIG.5. Effect of procaine hydrochloride on post-irradiation growth of fibrosarcoma in mice previously radio-sensitised by iothalamic acid. Dose — 3 krad for each irradiation.

sensitises only in anoxia [12]. However, the presence of these chemicals during irradiation was not fully effective in causing cell lethality as after about 8 days the tumours started regrowing at a rate similar to that of the control. Similar behaviour has been observed for higher doses also. As the presence of hypoxic and therefore radioresistant cells in tumours has been suggested as a possible cause of failure to eradicate the tumours [17], the double dose experiment was performed to see whether tumour recurrence could be prevented by a second dose of radiation after treatment with procaine, a hypoxic sensitiser. Though these experiments have not yet been completed, the preliminary results obtained so far indicate that the decrease in tumour size caused by irradiation in presence of procaine was practically the same as that caused by irradiation alone (Fig. 4). However, it is quite clear from Fig. 5 that irradiation in the presence of these sensitisers caused a significant

TABLE III. EFFECT OF IOTHALAMIC ACID AND PROCAINE HYDROCHLORIDE ON THE RADIOSENSITISATION OF MOUSE FIBROSARCOMA

Treatment	No. of mice	No. of mice cured[c]	Cure (%)
First and second irradiation after treatment with distilled water for each irradiation	36	7	19
First irradiation after treatment with ITA[a] and second irradiation after treatment with PRC[b]	32	15	47
First irradiation after treatment with ITA and second irradiation after treatment with distilled water	21	11	52
First irradiation after treatment with ITA and second irradiation after treatment with PRC	21	10	48

[a] ITA — iothalamic acid
[b] PRC — procaine hydrochloride
[c] Observation at the end of 85 days after irradiation. The dose delivered in each irradiation was 3 krad.

reduction in the tumour size when compared with the controls. This is further borne out by the observation that the cure of the animals increased from 19% in the controls to 47% when the irradiation was carried out in the presence of chemicals (Table III). On the other hand, no significant difference in the percentage cure could be observed when the second dose was given in the presence or absence of procaine — 48% and 52% respectively. Recently, a number of chemicals that specifically sensitise hypoxic cells have been developed [18—20]. Generally, these compounds have been chosen on the basis of their high electron affinity [6]. But the present compound procaine has been selected mainly because of its selective binding with membranes. In fact, a number of other membrane-specific compounds such as lignocaine, tetracaine, chlorpromazine etc. have been tested on bacterial systems and mammalian cells in vitro and found to sensitise anoxic cells [21]. Thus, it appears that membrane-specific drugs provide an additional group of compounds to be screened for potential application in clinical radiotherapy.

REFERENCES

[1] MULLENGER, L., SINGH, B.B., ORMEROD, M.G., DEAN, C.J., Nature (London) **216** (1967) 372.
[2] SHENOY, M.A., SINGH, B.B., GOPAL-AYENGAR, A.R., Science N.Y., **160** (1968) 999.
[3] SHENOY, M.A., JOSHI, D.S., SINGH, B.B., GOPAL-AYENGAR, A.R., Adv. Biol. Med. Phys. **13** (1970) 255.
[4] SINGH, B.B., SHENOY, M.A., GOPAL-AYENGAR, A.R., Ind. J. Exptl. Biol. **9** (1971) 518.
[5] MYERS, D.K., Int. J. Radiat. Biol. **19** (1971) 293.
[6] ADAMS, G.E., Br. Med. Bull. **29** (1973) 48.
[7] QUINTILIANI, M., in Advances in Chemical Radiosensitization (Proc. Panel, Stockholm, 1973), IAEA, Vienna (1973) 87.

[8] QUINTILIANI, M., MISITI-DORELLO, P., SAPORA, O., GEORGE, K.C., SHENOY, M.A., SINGH, B.B., Fifth Int. Cong. Radiat. Res., Seattle, USA (1974).
[9] MADDEN, R.E., BURK, D., J. Natl. Cancer Inst. 27 (1961) 841.
[10] SPURR, G.B., HORVATH, S.M., FERRAND, E.A., Am. J. Physiol. 186 (1956) 525.
[11] RITCHIE, J.M., COHEN, P.J., DRIPPS, R.D., in The Pharmacological Basis of Therapeutics (GOODMAN, L.S GILMAN, A., Eds), MacMillan, London (1970) 383.
[12] SHENOY, M.A., SINGH, B.B., GOPAL-AYENGAR, A.R., Nature (London) 248 (1974) 415.
[13] GEORGE, K.C., SHENOY, M.A., JOSHI, D.S., BHATT, B.Y., SINGH, B.B., GOPAL-AYENGAR, A.R., Br. J. Radiol. 48 (1975) 611.
[14] ALPER, T., in Biophysical Aspects of Radiation Quality (Proc. Symp. Lucas Heights, 1971), IAEA, Vienna (1971) 171.
[15] YATVIN, M.B., WOOD, P.G., BROWN, S.M., Biochim. Biophys. Acta 287 (1972) 390.
[16] NAIR, C.K.K., PRADHAN, D.S., Chem.-Biol. Interactions 11 (1975) 173.
[17] FOWLER, J.F., Clin. Radiol. 23 (1972) 257.
[18] FOSTER, J.L., WILLSON, R.L., Br. J. Radiol. 46 (1973) 234.
[19] ASQUITH, J.C., FOSTER, J.L., WILLSON, R.L., INGS, R., McFADZEAN, J.A., Br. J. Radiol. 47 (1974) 474
[20] BEGG, A.C., SHELDON, P.W., FOSTER, J.L., Br. J. Radiol. 47 (1974) 399.
[21] SHENOY, M.A., GEORGE, K.C., SINGH, B.B., GOPAL-AYENGAR, A.R., Int. J. Radiat. Biol. (in press).

APPROACHES TO SELECTIVE MODIFICATION OF RADIATION RESPONSE

E. RIKLIS, E. BEN-HUR
Israel Atomic Energy Commission,
Nuclear Research Center-Negev,
Beer-Sheva, Israel

Abstract

APPROACHES TO SELECTIVE MODIFICATION OF RADIATION RESPONSE.
　Selective modification of radiation response may be achieved by several factors which are often synergistically inter-related. The overall effect may then be: increase of damage and inhibition of the repair process (radiosensitization) versus decreased damage with enhanced repair process (radioprotection). Such factors may be biological – the enzymes of repair, chemical – polyamines as protectors, radiochemical – iodine-125 as sensitizer, or physical – temperature. The nature, magnitude and extent of the effects of some of these factors are discussed in detail. Haematoporphyrins, known to be selectively taken up by tumours and to increase their radiosensitivity, also cause increased dimerization of thymine. Polyamines, such as spermine, which are known to protect against ionizing radiation, also inhibit the photoreaction of psoralen and DNA. ^{125}I, when given as IUdR, causes cell killing more effectively than the equivalent amount of ^{3}H-thymidine. Hyperthermia is shown to interact synergistically with radiation in mammalian cells. It inhibits the repair of damage induced in DNA by various agents. The increased selective sensitivity of tumour cells to hyperthermia is of potential value in the combined treatment of radiation and heat in radiotherapy.

1. RADIATION DAMAGE AND ITS REPAIR

　Biological systems display a great variability in their response to u.v. or ionizing radiation, much of which is inherent and is in relation to their ability to repair or modify radiation damage. With u.v. irradiation, dimerization of adjacent pyrimidenes is generally considered the primary lethal and mutagenic lesion [1]. The effects of ionizing radiation are more diverse, including base damage in DNA (e.g. 5, 6 dihydroxy-dihydrothymine), single- and double-strand breaks [2].
　The cells are able to modify or ameliorate these lesions by a number of repair systems, first recognized in u.v.-irradiated bacteria. These are:

　Photoreactivation. Monomerization in situ of cyclobutyl pyrimidine dimers by DNA photolyase (photoreactivating enzyme) in a reaction requiring light of wavelength > 300 nm [3].

　Excision repair. This mode of repair involves the removal of damaged residues from the DNA by the concerted action of a number of enzymes. Although it was first demonstrated for u.v. irradiation [4–6], excision repair, unlike photoreactivation, operates on a whole spectrum of base damage as was suggested already in 1965 [6] and confirmed later by numerous studies (for review see Ref. [7]). A yet unpublished graph demonstrates proof of excision repair in vitro obtained in 1964 by Riklis [6] in *Escherichia coli* K-12 (Fig. 1). In-vitro studies [6, 8, 9] and the availability of radiation-sensitive bacterial mutants defective in specific repair enzymes elucidated many of the details of this repair mechanism, as shown in the scheme summarized by Grossman [10], who clarified much of the enzymatic system involved. The first step requires the recognition of the damage by an endonuclease followed by incision near the damage. The second step is exonucleolytic degradation of the damaged portion of the DNA and, as a result, removal of the damaged residue.

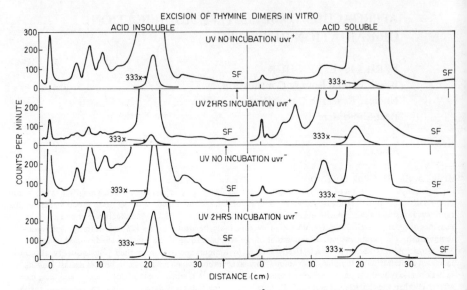

FIG.1. *Radiochromatogram of hydrolysates of u.v.-irradiated ^3H-thymidine-labelled DNA of* E. coli *demonstrating excision of thymine dimers in vitro by cell-free extracts of uvr$^+$ and not of uvr$^-$ strains. Excised dimers disappear from TCA-insoluble DNA and appear in acid-soluble fraction.*

In E. coli this is accomplished by the $5' \rightarrow 3'$ exonuclease associated with DNA polymerase I. This is followed by insertion of new bases into the single-strand gap, a process in which all three E. coli polymerases may participate. Repair is completed after opposed free ends are rejoined by polynucleotide ligase.

Post-replication repair. Also called recombinational repair, involves bypassing of damage in the DNA during semiconservative DNA synthesis and leaving single-strand gaps opposite the damaged residues [11]. These gaps are later filled with parental DNA by recombinational events in bacteria or by de novo DNA synthesis in mammalian cells [13]. This type of repair is not specific to u.v. damage [14]. Excision and post-replication repair occur in almost all types of cells which were examined, from mycoplasma to mammalian cells.

2. MODIFICATION OF RADIATION RESPONSE

Modifiers of radiation response, physical or chemical, act in two ways: by modification of the yield and nature of radiation products in DNA, and by interference with the repair of radiation damage. Among the first group, well known in ionizing radiation, are O_2 as radiosensitizer and thio compounds as radioprotectors. Inhibition of repair is displayed by caffeine, actinomycin D, quinacrine and many other chemicals. Acriflavine has a dual action, it reduces the yield of pyrimid dimers when present during u.v. irradiation, and inhibits repair when present after irradiation.

For a modifier of radiation response to be of practical value it must act differentially on norm and tumour cells so that there is an increase in therapeutic ratio. The only group of compounds shown to be effective up to now were the hypoxic sensitizers [15]. None of the repair inhibitors ha shown sufficient selective sensitization of tumour cells, as yet.

It is the purpose of this paper to present our results with respect to this problem, and to describe modifiers of radiation sensitivity which act either by affecting the type and magnitude of the lesion or by affecting the repair process. The variability in response to radiation due to the presence or absence of a repair system is of several orders of magnitude. This is far greater than any change inflicted by any modifier, whether sensitizer or protector, which acts directly on the site of damage by mechanisms such as radical scavenging, transfer of energy etc. It is clear that any modification in a repair system might, in theory, sensitize a cell by several orders of magnitude compared with only two- to three-folds obtained for example by cysteamine (for X-irradiation) or ~ 10-folds by acriflavine for u.v. light [16].

3. HAEMATOPORPHYRINS AS SENSITIZERS

Our techniques for determining u.v. photoproducts are well known and have been described [6, 17, 18]. We have been able to show that at least seven different identified products can be obtained upon u.v. irradiation of a frozen thymine solution, and four products upon irradiation of DNA in solution [19] and that the amount and relative composition of the products obtained depends on various physical factors such as dose, concentration, temperature and state of hydration.

Recent studies tie u.v. damage to malignant tumour development [20] and skin cancer to lack of ability of the cells to repair their damaged DNA [21], as is the case in Xeroderma pigmentosum patients. Yet, there are some new results (E. Friedberg, personal communication) indicating that in XP cells the lack of excision may be due to changes in the chromatin rather than due to lack in a repair enzyme. Chemicals which are naturally distributed in biological systems and can modify the amount and nature of the radiation or photoproducts are of special interest and possible therapeutic value.

Some 20 years ago, reports on increased sensitivity of paramecia to X-irradiation upon addition of haematoporphyrin (HP) were followed by reports on localisation of porphyrins in tumours, causing also a significant increase in their sensitivity to radiation. Thus, paramecia cells were shown to be 20 times more sensitive to X-irradiation when irradiated in the presence of haematoporphyrins [22], and tumour regression doubled when irradiated after injection of haematoporphyrins [23].

The fluorescence characteristic of porphyrins is thought to be possibly helpful in localization of tumours, such as by u.v. cistoscopy. It therefore seemed interesting to study the effect porphyrins may have on the formation of lesions, and more specifically the effect on the formation of thymine dimers. When a frozen solution of thymine has been irradiated in the presence of haematoporphyrin, a marked increase in the amount of thymine dimers, as expressed by the ratio of dimers to thymine, has been obtained [24].

Figure 2 shows graphically the products of u.v. irradiation of thymine. In addition to thymine dimers (T$\hat{T}$), in these experiments we have obtained an unidentified product marked B (possibly two products) which had disappeared when frozen thymine was irradiated in the presence of HP, while another product A appeared. Quantitatively it is shown that at a concentration of 10^{-5} M HP, 80% thymine dimers are produced (Table I). At optimal conditions of irradiation dose and HP concentration, namely 2000 ergs/mm^2 and 10^{-5} M HP, a three-fold increase in thymine dimer production is obtained (Fig. 3).

We have recently continued these experiments on DNA and whole bacterial cells where DNA was labelled with ^{3}H-thymidine. The results obtained (Riklis and Prager, to be published) show that at 1000 ergs/mm^2 the production of T$\hat{T}$ is doubled when HP is present in the irradiated solution. The nature of the mechanism of this "sensitization" is still unknown to us, but its possible implications are clear. HP also shows a slight absorption peak at 230 nm, and one possible mechanism for the HP effect is by enhanced dimer splitting.

In contrast, acridine orange has been shown to reduce the photoproduction of thymine dimers by a factor of 2 (Kramer and Riklis, unpublished results).

Recently, the use of phototherapy by shining light of 600 nm on certain tumours after injection of HP was shown to be effective in controlling tumour growth [25].

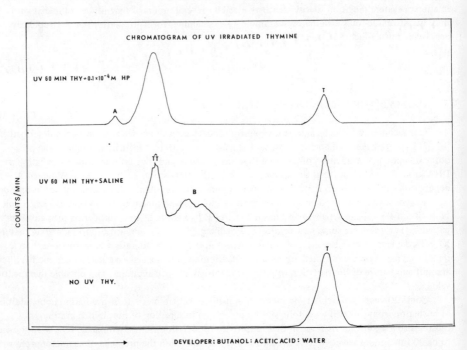

FIG.2. Radiochromatogram of irradiated thymine. ^{3}H-thymine was u.v.-irradiated in a frozen aqueous solution with a 15-W low-pressure u.v. lamp and a dose of 100 000 ergs/mm^2. The mixture was then chromatographed, developed in butanol:acetic acid:water and the paper strips then scanned for radioactivity.
Lower strip — control, not irradiated.
Middle strip — irradiated in saline solution
Top strip — irradiated with HP added to a final concentration of 0.1×10^{-4} M.

TABLE I. DEPENDENCE OF PHOTOPRODUCTS ON CONCENTRATION OF HAEMATOPORPHYRIN

HP Concn (M)	% Photoproducts			
	TÎ	T	a	b
0	38.9	34.4		26.5
0.05×10^{-4}	36.4	38.7		25.8
0.075×10^{-4}	56	26	3.0	15
0.1×10^{-4}	80	15	5	

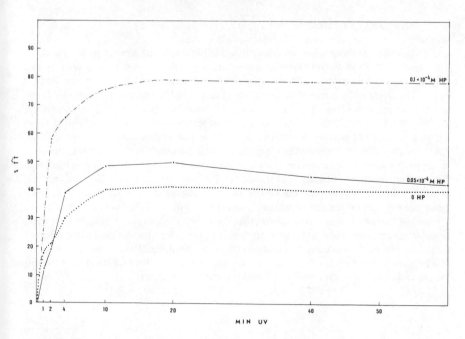

FIG.3. Ultra-violet dose effect on thymine in presence of haematoporphyrin. Irradiation conditions the same as those for Fig. 2. Dose-rate 20 ergs·mm^{-2}·s^{-1}.

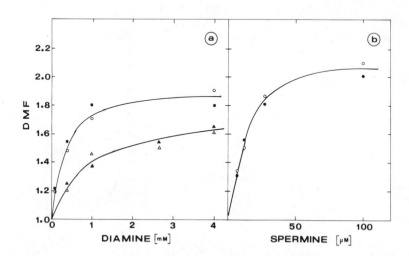

FIG.4. The dependence of the rate of the photoreactions of psoralen with DNA in 0.01M phosphate buffer at pH 6.8 on polyamine concentration. Closed and open symbols are for DNA cross-linking and psoralen addition to DNA respectively. (a) Circles and triangles are for the effects of cadaverine and diaminoethane respectively. (b) The effect of spermine.

4. PROTECTION BY POLYAMINES

Polyamines are compounds commonly found in living cells and serve as growth factors. They bind to DNA and stabilize its double-helix structure [26]. Because of their capacity to scavenge radicals they protect DNA from the effects of X-irradiation [27]. Since, for the DNA to photoreact, some distortion of the double-helix is required, we have studied the effect of polyamines on its photoreactivity with psoralen plus near u.v. irradiation. Figure 4 shows the dose-modifying factor (DMF) of three polyamines at various concentrations on the two photochemical reactions of psoralen with DNA, namely monoadduct production and cross-linking [28]. Evidently the effect on both reactions is the same. The magnitude of the effect is in the order spermine > cadaverine > diaminoethane which is also the same order for stabilizing the DNA [26]. Since under conditions of higher DNA stability (higher ionic strength) the effect was reduced, it is concluded that polyamines exert their action by stabilizing the DNA structure such that the distortion required for the photoreactions is made more difficult. Since both reactions are affected to the same extent it is concluded that the second reaction (the cross-linking of the complementary strand) does not require further distortion. We have also studied the effect of polyamines on the u.v.-induced dimerization of pyrimidines in the DNA. It appears to be much smaller than that of psoralen plus near u.v. The reason for this difference is not clear but might be due to differences in the extent of distortion required in each case.

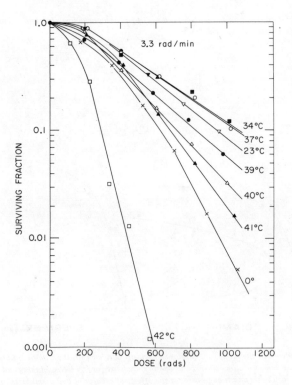

FIG.5. Survival of Chinese hamster cells X-irradiated at different temperatures.

5. ENHANCED RESPONSE OF HYPERTHERMIC MAMMALIAN CELLS TO RADIATION

X-irradiation

The interest in the possible application of local or systemic hyperthermia as a therapeutic modality in cancer treatment is of long duration and has been increasing during the last ten years (for review see Ref. [24]).

The recent observations that hyperthermia interacts synergistically with radiation [30–32] and drugs [33, 34] in cultured mammalian cells further emphasize the potential of hyperthermia. Hyperthermia seems to increase the therapeutic ratio when combined with radiation under certain conditions [35] which implies that tumour cells are selectively affected. This is due to the interaction of a number of factors. Most of the evidence suggests that tumour cells are more sensitive to heat than normal cells [36, 37]; hypoxic cells are at least as heat sensitive as oxygenated cells [38]; cells under poor nutritional conditions are more heat sensitive than cells that are properly nourished [39]. The importance of the last two factors is evident considering that the poorly nourished hypoxic cells in the tumour centre are the most radiation resistant.

Mammalian cells accumulate X-ray induced sublethal damage which they are able to repair during post-irradiation incubation (for review see Ref. [40]). The effect of hyperthermia during irradiation at a low dose-rate is shown in Fig. 5. The increased survival when irradiation is at 37°C compared with 0°C is due to repair of sublethal damage during irradiation. As the temperature is increased above 37°C the cells become progressively more sensitive. This effect is inversely related to the dose-rate as Fig. 6 shows. Here the magnitude of the effect is plotted against the reciprocal of the temperature (Arrhenius plots) for three dose-rates. Apparently, the synergism is not linearly dependent on temperature. This indicates that a single activation energy does not apply.

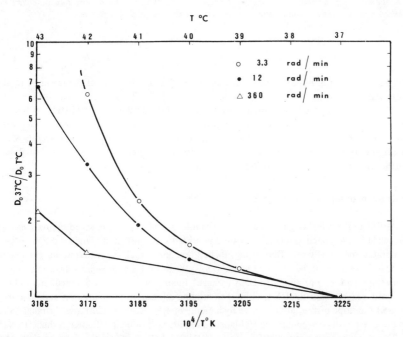

FIG.6. *Arrhenius plots of the ratios of the survival-curve slopes (D_0) of Chinese hamster cells X-irradiated at various temperatures with a dose-rate of 3.3 rad/min, ○; 12 rad/min, ●; 360 rad/min, △.*

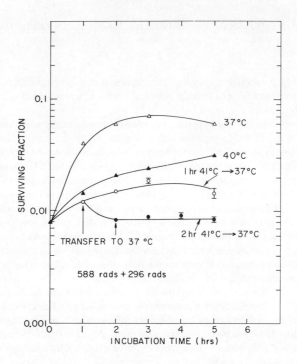

FIG. 7. *Two-dose fractionation survival of Chinese hamster cells. Cells were given a first dose of 588 rad, incubated for various times at different temperatures and then exposed to a second dose of 296 rad.*

That the effect of hyperthermia is at least partly due to inhibition of the repair of sublethal damage was demonstrated by fractionated exposures (Fig. 7). There is an apparent slower recovery at 40°C which is completely abolished at 41°C. This inhibition is not immediately reversed upon transfer to 37°C. Further evidence for inhibition of the repair of sublethal damage is indicated by fractionated survival curves (Fig. 8). The displacement of the survival curve downwards when incubation is at 41.5°C indicates that, in addition to interference with repair of sublethal damage, hyperthermia also enhances expression of lethal damage.

Suicide by incorporation of radionuclides

Although DNA is thought to be the target molecule when cells are killed by ionizing radiation, other cell structures such as membranes are undoubtedly also affected. When radionuclides are incorporated specifically into DNA using labelled precursors such as ^{3}H-thymidine, the radiation damage is more likely to be confined to the DNA. We used this approach to show that the effect of hyperthermia is due primarily to DNA damage. Figure 9a shows that 5-h incubation at 42°C after unifilar labelling of the DNA with ^{3}H-thymidine essentially eliminated the shoulder and decreased the slope of the survival curve by a factor of 1.8. Results qualitatively similar to these were obtained with cells containing ^{125}I-labelled iododeoxyuridine (^{125}IUdR) in their DNA (Fig. 9b).

There are, however, several differences. ^{125}IUdR-labelled cells survive exponentially already at 37°C and incubation at 42°C decreases D_0 only by a factor of 1.2. As found with frozen cells

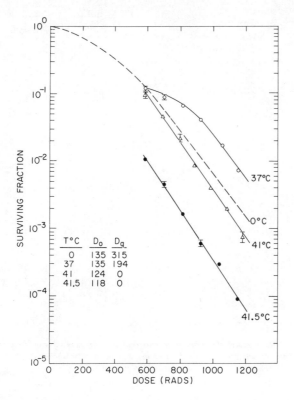

FIG.8. *Fractionation survival curves of Chinese hamster cells. Cells were exposed to a conditioning dose of 593 rad at 0°C, incubated for 2.5 h at 37°C, ○; 41°C, △; and 41.5°C, ●. They were then exposed to graded second doses at 0°C. The dashed line is the single-dose survival curve.*

[41], we also observe under growth conditions that ^{125}I in DNA is much more toxic to the cells than ^{3}H, about 9 times fewer disintegrations being required to kill a cell (D_0 at 37°C = 0.031 dis per min per cell and 0.27 dis per min per cell, respectively).

Quintiliani [42] suggested the use of iodine compounds as radiosensitizers affecting various cell systems, mostly protein. We are suggesting here the use of an iodine compound which can act directly on DNA.

DNA damage and repair

The effect of hyperthermia on the repair of X-ray-induced scissions of DNA single-strands and damage to a DNA complex was also studied using the alkaline sucrose gradient technique [43], as modified by Elkind and Kamper [44] and Elkind [45]. Figure 10a shows that single-strand breaks induced by 10 krad are rejoined at 37°C within 5 h. At 42°C there is initial rejoining which, however, is not completed (Fig. 10b). Instead, DNA degradation commences after about 2 h.

Under gentle lysis conditions, a DNA structure is observed which was termed the "DNA complex" [44]. The complex is much more sensitive to radiation than is the single-stranded DNA, probably because its size is much larger. Thus, the effects of small X-ray doses can be studied by

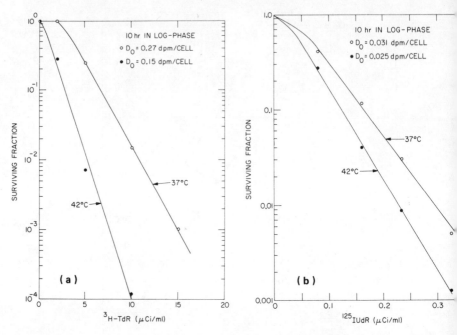

FIG.9. *Temperature effect on survival of Chinese hamster cells unifilarly labelled with ^{3}H-thymidine or ^{125}IUdR. Survival of Chinese hamster cells grown at 37°C for 10 h in log phase (unifilar labelling) with (a) ^{3}H-thymidine and (b) ^{125}I-iododeoxyuridine and then incubated for 5 h at 37°C (○) and 42°C (●).*

making use of the DNA complex. Figure 11a shows that irradiation at 37°C with 12 rad/min (total dose 1600 rad) has no significant effect on the complex, although the same dose at 0°C breaks the complex apart. This is because at 37°C the rate of rejoining exceeds the rate of breakage induced with 12 rad/min. A similar exposure at 42°C causes most of the complex to disappear (Fig. 11b). This implies that under these conditions the rejoining process is slowed down. It is not completely inhibited since 1600 rad without repair result in the complete disappearance of the complex and the appearance of single-stranded DNA as a well-defined peak at the top of the tube [

The relevance of such studies to cell survival was indicated by earlier work of Kaplan [47, 48].

DISCUSSION

Approaches for selective modification of radiation response are presented. One takes advantage of the observation that tumours take up certain drugs selectively, in this case haematoporphyrins. Then, if it can be shown that this drug can increase the radiation response of cells, an increase in the therapeutic ratio will probably result. We have attacked the problem from the photochemical point of view, first showing enhanced photodimarization of thymine in frozen solution in the presence of haematoporphyrins. We then proceeded to more complex systems, namely DNA and bacteria. The preliminary results suggest enhanced response under certain conditions. The next obvious step will be to show that this drug also enhances response to ionizing radiation and to establish, by in-vivo experiments, what the increase in therapeutic ratio is and what are the optimal conditions for a maximal effect.

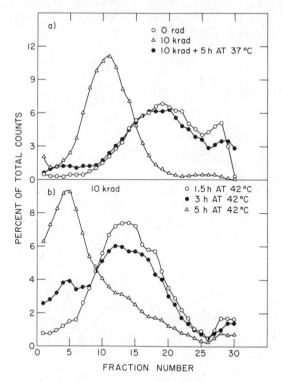

FIG.10. Temperature effect on sedimentation pattern of X-irradiated DNA. Sedimentation in alkaline sucrose gradients of DNA from Chinese hamster cells. Cells, uniformly labelled with ^{3}H-thymidine, were lysed for 60 min at 42°C in a 0.25-ml top layer (0.45N NaOH, 0.5M NaCl, 0.01M Na_2EDTA) on top of 5–20% alkaline sucrose gradients (0.3N NaOH, 0.7M NaCl and 0.003M Na_2EDTA). Sedimentation was at 11000 rev/min for 17.1 h at 3°C in a Beckman SW 50.1 rator. (a) Unirradiated cells, ○; cells exposed to 10 krad at 0°C, △; cells incubated for 5 h at 37°C after exposure to 10 krad at 0°C, ●. (b) Cells exposed to 10 krad at 0°C and then incubated at 42°C for 1.5 h, ○; 3 h, ●; 5 h, △. Sedimentation from left to right.

Another approach makes use of the intrinsic higher sensitivity of cancer cells to hyperthermia. Although much work was done using heat alone as a modality in the treatment of cancer patients, success was very limited. We have shown that hyperthermia interacts synergistically with radiation. It is shown that the effect is due partly to inhibition of the repair of sublethal damage and partly to enhanced expression of lethal damage. Since hyperthermia interacts synergistically with agents which kill cells by causing DNA damage, the involvement of DNA repair was studied. It was found that hyperthermia interferes with the repair of single-strand breaks in DNA and that DNA degradation is the end result.

We suggest therefore that the inhibition of cellular recovery could be due to the interference with the repair of DNA damage while the enhanced expression of lethal damage may be due to the enhanced DNA degradation. Practically, the combined modality of radiation and hyperthermia could be more effective in tumour control not only because of their synergism but also because hyperthermia tends to kill more efficiently those cells in the tumours that are the most radiation resistant, namely poorly oxygenated and poorly nourished cells in the tumour centre.

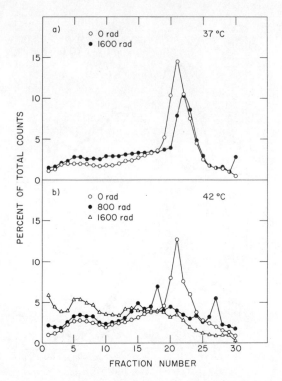

FIG.11. Temperature effect on sedimentation pattern of X-irradiated "DNA complex". Sedimentation in alkaline sucrose gradients of DNA from Chinese hamster cells. Gradients and top layer as in Fig. 7. Lysis was for 10 min at 25°C and sedimentation was at 5000 rev/min for 16.25 h at 0°C. Cells were irradiated with 12 rad/min at the indicated temperatures.

It is also shown that naturally occurring compounds such as polyamines can modify the photoreactivity of DNA and thereby may affect the survival response of the cells. If we could change the concentration of these compounds within the cells we would be able to control their radiation response. We suggest that polyamines could be one of the factors affecting DNA photochemistry in vivo. Thus, some of the variation in the sensitivity of Chinese hamster cells to psoralen plus light as a function of their position in the cell cycle [49] could be explained on this basis. Also, polyamines may contribute to the decreased rate to photodimerization of pyrimidines observed in vivo [50].

Finally, we would like to emphasize our belief that, in view of the inherent potential of repair modifiers to affect considerably the response of cells to radiations, it is desirable to continue work in this direction aimed at obtaining sensitizers and protectors which may act selectively to increase the differential response in such a way that will increase the therapeutic ratio.

ACKNOWLEDGEMENTS

The authors wish to acknowledge with thanks the contributions made to the work described here by their respective collaborators: Mrs. A. Prager, Mrs. E. Elhanani, Dr. M.M. Elkind and Dr. B.V. Bronk.

REFERENCES

[1] SETLOW, R.B., Cyclobutane-type pyrimidine dimers in polynucleotides, Science 153 (1966) 379.
[2] SETLOW, R.B., SETLOW, J.K., Effects of radiation on polynucleotides, Annu. Rev. Biophys. Bioeng. 1 (1972) 293.
[3] RUPERT, C.S., HARM, W., HARM, H., "Photoenzymatic repair of DNA. II Physical/chemical characterization of the process", Molecular and Cellular repair processes (BEERS, R.F., HERRIOT, R.M., TILGHMAN, R.C., Eds), The Johns Hopkins Med. J., Suppl. 1 (1972) 64.
[4] SETLOW, R.B., CARRIER, W.L., The disappearance of thymine dimers from DNA: An error correcting mechanism, Proc. Natl. Acad. Sci. U.S.A. 51 (1964) 226.
[5] BOYCE, R.P., HOWARD-FLANDERS, P., Release of ultraviolet light induced thymine dimers from DNA *Escherichia coli* K-12, Proc. Natl. Acad. Sci. U.S.A. 51 (1964) 293.
[6] RIKLIS, E., Studies on the mechanism of repair of ultraviolet-irradiated viral and bacterial DNA in vivo and in vitro, Can. J. Biochem. 43 (1965) 1207.
[7] CERUTTI, P.A., Excision repair of DNA base damage, Life Sci. 15 (1974) 1567.
[8] CARRIER, W.L., SETLOW, R.B., Excision of pyrimidine dimers from irradiated deoxyribonucleic acid in vitro, Biochim. Biophys. Acta 129 (1966) 318.
[9] RORSCH, A., VAN DER KAMP, C., ADEMA, J., Dark reactivation of ultraviolet irradiated bacteriophage DNA in vitro, Biochim. Biophys. Acta 80 (1964) 346.
[10] GROSSMAN, L., Enzymes involved in the repair of DNA, Adv. Radiat. Biol. 4 (1974) 77.
[11] RUPP, W.D., HOWARD-FLANDERS, P., Discontinuities in the DNA synthesized in an excision defective strain of *Escherichia coli* following ultraviolet irradiation, J. Mol. Biol. 31 (1968) 291.
[12] RUPP, W.D., WILDE, C.E., RENO, D.L., HOWARD-FLANDERS, P., Exchanges between DNA strand in ultraviolet-irradiated *Escherichia coli*, J. Mol. Biol. 61 (1971) 25.
[13] LEHMANN, A.R., Postreplication repair of DNA in ultraviolet-irradiated mammalian cells, J. Mol. Biol. 66 (1972) 319.
[14] BEN-HUR, E., ELKIND, M.M., Postreplication repair of DNA containing psoralen addition products in Chinese hamster cells, Proc. XI Int. Cancer Cong., Florence, Italy (1974).
[15] ADAMS, G.E., ASQUITH, J.C., WATTS, M.E., SMITHEN, C.E., Radiosensitization of hypoxic cells in vitro: a water-soluble derivative of paranitroacetophenone, Nature (London) New Biol. 239 (1972) 23.
[16] FORAGE, A.G., PORTEN-SEIGNE, I., Thymine dimerization in ultraviolet-irradiated bacteria treated with acriflavine, Photochem. Photobiol. 20 (1974) 81.
[17] KABANTCHICK, Y., RIKLIS, E., Photoproducts of ultraviolet irradiated thymine and DNA, Israel J. Chem. 6 (1968) 102.
[18] KRAMER, J., RIKLIS, E., Photoproducts formation in UV irradiated DNA at high temperatures and high irradiation doses, Intl. J. Radiat. Biol. 23 (1973) 75.
[19] RIKLIS, E., "Photoproducts and their significance in radiation damage repair", Int. Symp. New Trends in Photobiology (CALDAS, L.R., Ed.), An. Acad. Bras. Cienc. 45 (1973) 221.
[20] SETLOW, R.B., "The relevance of photobiological repair", Int. Symp. New Trends in Photobiology (CALDAS, L.R., Ed.), An. Acad. Bras. Cienc. 45 (1973) 215.
[21] CLEAVER, J.E., Defective repair replication of DNA in xeroderma pigmentosum, Nature (London) 218 (1968) 652.
[22] FIGGE, F.H.J., WICHTERMAN, R., Effect of hematoporphyrin on X-irradiation sensitivity in paramecium, Science 122 (1955) 468.
[23] SCHWARTZ, S., ABSOLON, K., VERMUND, H., Effect of porphyrins on X-ray sensitivity of tumors, J. Lab. Clin. Med. 46 (1955) 949.
[24] RIKLIS, E., PRAGER, A., ELHANANI, E., Effects of hematoporphyrins on photoproduct formation in UV-irradiated thymine, Proc. 5th Int. Congr. Radiat. Res., Seattle (1974) 193.
[25] DOUGHERTY, T.J., GRINDEY, G.B., FIEL, R., WEISHAUPT, K.R., BOYLE, D.G., Photoradiation therapy II. Cure of animal tumors with hematoporphyrins and light, J. Natl. Cancer Inst. 55 (1975) 115.
[26] MAHLER, H.R., MEHROTA, B.D., The interaction of nucleic acids with diamines, Biochim. Biophys. Acta 68 (1963) 211.
[27] BROWN, P.E., Some effects of binding agents on the X-irradiation of DNA, Radiat. Res. 34 (1968) 24.
[28] BEN-HUR, E., ELKIND, M.M., DNA crosslinking in Chinese hamster cells exposed to near ultraviolet light in the presence of 4, 5', 8-trimethyl-psoralen, Biochim. Biophys. Acta 331 (1973) 181.
[29] SUIT, H.D., SHWAYDER, M., Hyperthermia: potential as an anti-tumor agent, Cancer 34 (1974) 122.
[30] BEN-HUR, E., BRONK, B.V., ELKIND, M.M., Thermally enhanced radiosensitivity of cultured Chinese hamster cells, Nature (London) New Biol. 238 (1972) 209.

[31] BEN-HUR, E., ELKIND, M.M., BRONK, B.V., Thermally enhanced radioresponse of cultured Chinese hamster cells: inhibition of repair of sublethal damage and enhancement of lethal damage, Radiat. Res. 58 (1974) 38.
[32] ROBINSON, J.E., WIZENBERG, M.J., Thermal sensitivity and the effect of elevated temperatures on the radiation sensitivity of Chinese hamster cells, Acta. Radiol. 13 (1974) 241.
[33] BEN-HUR, E., ELKIND, M.M., Thermal sensitivity of Chinese hamster cells to methyl methane-sulfonate: relation of DNA damage and repair to survival response, Cancer Biochem. Biophys. 1 (1974) 23.
[34] HAHN, G.M., BRAUN, J., HAR-KEDAR, I., Thermochemotherapy: synergism between hyperthermia (42–43°C) and adriamycin (or bleomycin) in mammalian cell inactivation, Proc. Natl. Acad. Sci. U.S.A. 72 (1975) 937.
[35] THRALL, D.E., GILLETTE, E.L., DEWEY, W.C., Effect of heat and ionizing radiation on normal and neoplastic tissue of the C3H mouse, Radiat. Res. 63 (1975) 363.
[36] CAVALEIRE, R., CIOCATTO, E.C., GIOVANELLA, B.C., HEIDELBERGER, C., JOHNSON, R.O., MARGOTTINI, M., MONDOVI, B., MORICCA, G., ROSSI-FANELLI, A., Selective heat sensitivity of cancer cells, Biochemical and clinical studies, Cancer 20 (1967) 1351.
[37] KASE, K., HAHN, G.M., Different heat response of normal and transformed human cells in tissue culture, Nature (London) 255 (1975) 228.
[38] GERWECK, L.E., GILLETTE, E.L., DEWEY, W.C., Killing of Chinese hamster cells in vitro by heating under hypoxic or aerobic conditions, Europ. J. Cancer 10 (1974) 691.
[39] HAHN, G.M., Metabolic aspects of the role of hyperthermia in mammalian cell inactivation and their possible relevance to cancer treatment, Cancer Res. 34 (1974) 3117.
[40] ELKIND, M.M., WHITMORE, G.F., The Radiobiology of Cultured Mammalian Cells, Gordon and Breach, New York (1967).
[41] BURKI, H.J., ROOTS, R., FEINENDEGEN, L.E., BOND, V.P., Inactivation of mammalian cells after distintegrations of ^{3}H or ^{125}I in cell DNA at $-196°$C, Int. J. Radiat. Biol. 24 (1973) 363.
[42] QUINTILIANI, M., "Molecular mechanism of radiosensitization by iodine-containing compounds", Advances in Chemical Radiosensitization (Proc. Panel Stockholm, 1973), IAEA, Vienna (1974) 87.
[43] McGRATH, R.A., WILLIAMS, R.W., Reconstruction in vivo of irradiated *Esherichia coli* DNA: The repairing of broken pieces, Nature (London) 212 (1966) 534.
[44] ELKIND, M.M., KAMPER, C., Two forms of repair of DNA in mammalian cells following irradiation, Biophys. J. 10 (1970) 237.
[45] ELKIND, M.M., Sedimentation of DNA released from Chinese hamster cells, Biophys. J. 11 (1971) 502.
[46] BEN-HUR, E., ELKIND, M.M., Thermally enhanced radioresponse of cultured Chinese hamster cells: damage and repair of single-strand DNA and a DNA complex, Radiat. Res. 59 (1974) 484.
[47] KAPLAN, H.S., DNA strand scission and loss of viability after X-irradiation of normal and sensitized bacterial cells, Proc. Natl. Acad. Sci. U.S.A. 55 (1966) 1442.
[48] KAPLAN, H.S., "Repair of X-ray damage to bacterial DNA and its inhibition by chemicals", Advances in Chemical Radiosensitization (Proc. Panel Stockholm, 1973), IAEA, Vienna (1974) 123.
[49] BEN-HUR, E., ELKIND, M.M., Psoralen plus near ultraviolet light inactivation of cultured Chinese hamster cells and its relation to DNA cross-links, Mutation Res. 18 (1973) 315.
[50] UNRAU, P., WHEATCROFT, R., COX, B., OLIVE, T., The formation of pyrimidine dimers in the DNA of fungi and bacteria, Biochim. Biophys. Acta 312 (1973) 626.

RADIATION-DOSE DEPENDENCE OF SENSITIZATION BY ELECTRON-AFFINIC COMPOUNDS[*]

L. RÉVÉSZ, B. LITTBRAND
Department of Tumor Biology,
Karolinska Institute Medical School,
Stockholm, Sweden

Abstract

RADIATION-DOSE DEPENDENCE OF SENSITIZATION BY ELECTRON-AFFINIC COMPOUNDS.
   In an extension of previous studies, Chinese hamster cells were treated with the sensitizing compound Ro 07-0582 and irradiated under extremely hypoxic conditions. In one series of experiments, the survival ratios of the sensitizer-treated to untreated cells were determined after exposure to different radiation doses up to 400 R; in a complementary series, pairs of radiation doses which gave equal survival for the sensitizer-treated and untreated cells in the range 0–400 R were determined; in a third series, the survival parameters of the cells exposed to doses of 400 R and larger were calculated. Analysis of the data suggested that, in the large dose region, sensitization with a dose-modifying factor of about 2 occurs after treatment with the compound at a concentration of 8mM; in the low dose region, sensitization after treatment of the cells with a similar concentration of the compound is absent or considerably decreased; treatment with the compound at a low concentration (0.2mM) may be protective. The results are discussed in relation to previous observations on the sensitizing effect of oxygen and other electron-affinic sensitizers in various concentrations. The data are also considered from a practical point of view for a three-compartment tumour model consisting of anoxic, hypoxic and fully oxic cell populations.

INTRODUCTION

   The IAEA panel organized in Stockholm in 1973 expressed an optimistic view regarding the potential value of electron-affinic sensitizers in radiotherapy [1]. Experimental results which have accumulated during the past two years strongly support this view.
   Many of the advances have been summarized in an excellent review article [2] which also discusses some recent relevant questions. Among the current problems, those of particular importance concern the preliminary clinical trials which were started with some electron-affinic drugs [3, 4]. It has been pointed out [5, 6] that the efficacy of such drugs may be greatly influenced by the regime of dose fractionation used in the radiotherapy. The results of clinical trials with radiotherapy in combination with hyperbaric oxygen as a sensitizer indicate [7–9] that the effectiveness of this type of treatment is related to the use of a small number of large dose fractions. Data from model experiments summarized in our report to the panel in 1973 [10] suggest that a modification of the conventional fractionation may be necessary also for the effective use of the electron-affinic chemical compounds that mimic oxygen.
   This question is discussed here in regard to recent observations we have made in experiments with a nitro-imidazole, Ro 07-0582. In extensive tests in different laboratories, this compound has been found to be an effective sensitizer of hypoxic cells [2].

---

[*] This work was supported by grants from the Swedish Cancer Society.

CHANGES IN SURVIVAL CURVE PARAMETERS

In one series of replicate experiments, the effect of the compound Ro 07-0582 at a concentration of 8mM was tested on the radiation survival of Chinese hamster cell cultures. Irradiation was done under severe hypoxic conditions in argon with less than 3 ppm oxygen impurity according to a method described in earlier publications [11, 12]. Radiation doses were used which decreased the surviving fraction to 10% or less. In considering the multitarget single-hit formula, statistical least-square analysis of the survival of the untreated and treated cultures gave D_0 values of 267 R and 123 R, and extrapolation numbers (n) of 1.4 and 2.3, respectively [13].

The finding that the drug decreases D_0 and increases n agrees well with observations we and another group made with some other electron-affinic compounds [14, 15]. These changes of the survival parameters imply that the radiation-dose modifying effect of the drug is not constant, but that it varies at different survival levels [13]. When either a single-component or a two-component multitarget single-hit model is considered, the effect of the drug in increasing n and decreasing D_0 results in survival curves which cross each other. At the crossing point, the dose-modifying factor (DMF) is unity, i.e. neither sensitization nor protection occurs. With radiation doses larger than the one which corresponds to the crossing point, the DMF expressing sensitization gradually increases and reaches, at large doses, a maximum defined by the ratio of the D_0 of the drug-treated cells to that of the untreated cells. With radiation doses smaller than that corresponding to the crossing point, DMF becomes smaller than unity, thus indicating protection instead of sensitization. The Chadwick model, which describes the radiation survival curves with a linear and quadratic term in dose corresponding to a single- and two-event mechanism, also gives crossing survival curves if the constant of the linear term is increased while that of the quadratic term approaches zero.

From the D_0 of the sensitized and untreated cultures already indicated, the ratio 2.17 can be calculated as the maximum DMF for the sensitizer at a concentration of 8mM. The prediction that DMF approaches unity, or becomes even smaller at large survival levels after irradiation with low doses, was tested in experiments in which the cells were exposed to radiation in a dose range corresponding to the "shoulder region" of the survival curves.

DOSE-MODIFYING FACTORS IN THE SHOULDER REGION

Two types of experiments were conducted. In one of them, to which reference has already been made in a preliminary report [13], the survival ratio of the sensitizer-treated (8mM) to untreated cells was determined after practically anoxic exposure to different doses in the range 50–200 R.

TABLE I. DOSE-MODIFYING FACTORS IN A RANGE OF LOW RADIATION DOSES

Data from experiments presented in Figs 1 and 2 were used for the calculations

Sensitizer concentration (mM)	Exposure dose to sensitizer-treated cells (R)	Dose-modifying factor		
		Mean	Range defined by the survival variation of:	
			Untreated controls	Sensitizer-treated cells
8	100	1.17	0.92–1.42	0.75–1.62
8	200	1.54	1.44–1.66	1.37–1.72
0.2	200	1.17	0.98–1.42	0.97–1.33
0.2	400	0.85	0.81–0.90	0.80–0.90

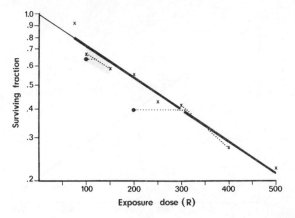

Fig.1. Radiation survival of Chinese hamster cells untreated (x) and pretreated (●) with 8mM Ro 07-0582 before exposure in severe hypoxia (< 3 ppm oxygen). Dose-modifying factors are indicated by extrapolation from the two doses which reduced the surviving fraction of the untreated controls closest above and below the surviving fraction of the pretreated cells (broken lines). Shaded areas show survival ranges defined by the 95% confidence limits of the surviving fraction of the relevant controls. The solid line illustrates the survival regression of the controls determined by least-square analysis (n = 0.99, D_0 = 325 R). Means calculated from nine replicate experiments are presented.

Statistical analysis of the ratios indicated that they did not differ to any significant extent from unity. While this result is consistent with the prediction, the lack of a statistically significant difference is obviously no proof of identity, and the data cannot be regarded as evidence of a total loss of sensitization in the dose range tested. However, they strongly suggest a radiation-dose dependence of the sensitizing effect.

In this regard, results obtained in another type of experiment are more conclusive since they permit the direct calculation of DMF by which sensitization occurs at different survival levels. In these experiments, the survival of the sensitizer-treated cell cultures was calculated after exposure to a certain predetermined radiation dose that was delivered under practically anoxic conditions. Untreated control cultures were irradiated with a series of different doses giving survivals similar to those of the sensitized cultures. A more accurate estimate of this dose was made by extrapolation from the two doses which reduced the surviving fraction of the controls to a value closest, above and below, the surviving fraction of the sensitized cells. The mean surviving fractions determined in replicate experiments were used for the calculation of the mean DMF. We calculated two different ranges for this: one defined by the 95% confidence limits of the surviving fraction of the controls and the other defined by that of the sensitizer-treated cells. It can be expected that the two ranges will largely overlap (see Table I) and that the true DMF will lie within these ranges.

Figure 1 illustrates the result of the experiments in which the sensitizer was used at a concentration of 8mM. At about the 65% survival level, sensitization is clearly reduced compared with that seen at about the 40% level. At the larger survival level, DMF is close to unity (Table I). At the lower survival level, DMF is larger than unity but, considered even in its widest statistical range, it is still smaller than the D_0 ratio 2.17 referred to previously as the sensitizing factor for survival levels far below 10% and determined similarly with the 8mM sensitizer.

Although these results clearly indicate the predicted decrease in the extent of sensitization for a range of low radiation doses and agree with some other, recent observations [16], they do not suggest that any significant protection occurs in this range. The lack of protection can be explained by considering the large decrease of the D_0 by the sensitizer at the particular concentration used.

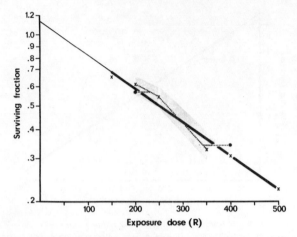

FIG.2. Same as for Fig.1 but with the sensitizer at a concentration of 0.2mM. Survival parameters of the controls: $n = 1.14$, $D_0 = 305$ R. Means calculated from eight replicate experiments are illustrated.

TABLE II. SURVIVAL RATIO OF SENSITIZER-TREATED TO UNTREATED CELLS IN A RANGE OF LOW RADIATION DOSES

Data from experiments presented in Figs 1 and 2 were used for the calculations

Sensitizer concentration (mM)	Exposure dose (R)	Survival ratio of sensitizer-treated cells to untreated cells
8	100	$0.95 \begin{smallmatrix} +0.05 \\ -0.04 \end{smallmatrix}$
8	200	$0.71 \begin{smallmatrix} +0.05 \\ -0.04 \end{smallmatrix}$
0.2	200	$0.93 \begin{smallmatrix} +0.06 \\ -0.06 \end{smallmatrix}$
0.2	400	$1.08 \begin{smallmatrix} +0.04 \\ -0.03 \end{smallmatrix}$

It has been pointed out [13] that, as D_0 decreases, the crossing point moves in the direction of low radiation doses, and possible protection may therefore become less detectable. As has also been shown experimentally with oxygen in low concentrations [12], the chances to detect any protective effect are improved with the crossing point moved towards larger doses, i.e. when D_0 remains high due to a less efficient sensitization. Therefore, in a complementary test, the sensitizer was used at a 40-times lower concentration at which, according to recent observations [17], it still retains a considerable sensitizing effect. As indicated in Fig. 2, DMF is again close to unity at about the 60% survival level. At about the 35% level, however, it is smaller than unity even considering the widest statistical range, and indicates a protecting instead of a sensitizing effect.

These experiments were designed primarily to determine DMF at different survival levels. In addition, the data permit the calculation of the survival ratio of the sensitizer-treated to untreated cells in the same way as in the other type of experiment already referred to which was designed specifically for this purpose. The ratios listed in Table II are clearly consistent with the comparable DMF values. Thus, they do not differ significantly from unity when the exposure dose is 100 R and 200 R and the sensitizer concentration is 8mM and 0.2mM, respectively. When the exposure doses are increased, the survival ratio becomes significantly smaller than unity and indicates sensitization if the drug is present at a concentration of 8mM; and significantly larger than unity and indicates protection if present at a concentration of 0.2mM.

ROLE OF OXYGEN CONTAMINATION

To detect any rise in extrapolation number by the sensitizer, severe hypoxic conditions are essential. These should decrease the number to unity as has been found previously [11, 12] and is also indicated by the survival curve of the controls in Figs 1 and 2. It has been demonstrated earlier [12] that the extrapolation number may not be depressed to this value if oxygen is present even at concentrations as low as 100–200 ppm. Under such conditions, solely the slope of the survival curve will be changed by the sensitizer, i.e. the effect may appear simply as dose modifying by a constant factor at all survival levels. This can explain many of the conclusions reached in experiments in which the oxygen contamination of the anoxic system may not have been controlled to a sufficient extent. Penetration of oxygen through rubber tubes, dissolved oxygen in plastics, insufficient deoxygenation of the cellular medium, back-flow of air in the system in the absence of a lock for the outflowing nitrogen or argon, or the oxygen impurity of the gases themselves can be considered, amont other factors, as possible sources of contamination which require special controls.

The role of some of these factors has been clearly demonstrated in a "didactic" experiment in which nitrogen or argon of similar high purity, with an oxygen content of less than 3 ppm, was used to create hypoxic conditions for the cells during irradiation. The same experimental system and conditions, developed earlier in our laboratory for exposures in severe hypoxia, were used in both cases with the exception that nitrogen, in contrast to argon, was led to the cells in the irradiation chamber through a 55-cm long rubber tube (2 mm ϕ) instead of metallic tubing and that, after passing the irradiation chamber, it was led into the air without any water lock. When measured after leaving the chamber, no increased oxygen contamination was detected in argon. On the other hand, the insertion of the rubber tube raised the oxygen content of the nitrogen to about 50 ppm. Since the oxygen-measuring apparatus itself may function as a lock for the outflowing gas and prevent the back-flow of air, the actual oxygen contamination of the nitrogen atmosphere inside the chamber was probably larger than 50 ppm.

Figure 3 shows that the survival curve of the cells exposed in argon is nearly exponential ($n = 1.08$, $D_0 = 279$). In contrast, the survival curve of the cells irradiated in nitrogen has an increased extrapolation number and a considerable shoulder region, while the slope shows only a slight change ($n = 2.12$, $D_0 = 240$). Besides the oxygen contamination of nitrogen, the difference can be attributed, to some extent, to a difference in some physiological effect between nitrogen and argon. It has been suggested that noble gases hinder the passage of oxygen into cells [18].

THE THREE-COMPARTMENT MODEL

The practical relevance of the observations described here will greatly depend upon the answer to the question whether cells do exist in tumours under the severely hypoxic conditions used in our experiments. On theoretical grounds we cannot, at present, exclude this possibility which indeed finds support in some indirect experimental observations. Thus, different mammalian cell types

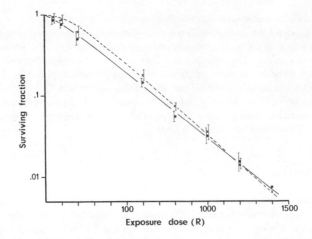

FIG.3. Radiation survival curve of Chinese hamster cells exposed under hypoxic conditions created with argon (●——●) or nitrogen (x - - - x) with <3 ppm oxygen impurity. Nitrogen was led to the irradiation chamber through rubber tubes and left it without passing through a lock; the argon passed through metallic tubing and a water lock. Lines illustrate survival curves calculated according to the multi-target single-hit model with constants determined by least-square analysis of the survival data below the 10% level. The means and their 95% confidence limits are shown for six and seven replicate experiments with argon and nitrogen respectively.

have been shown to survive in culture for prolonged periods of time under conditions practically deficient of oxygen [19, 20]. In experimental tumours a chronically hypoxic cell population has been demonstrated to exist with a decreased capacity to repair radiation damage [21], a property indicating severe hypoxia in the cellular microenvironment.

In considering the oxygen gradient in tumours and the oxygen-dependent variation of the cellular radiation response, a previous paper [22] proposed a three-compartment model for the cell population in tumours: one compartment comprising anoxic cells resistant to radiation and deficien in the capacity to repair radiation damage; another comprising hypoxic cells also resistant to radiatio but capable of repairing damage; and a third one with fully oxic cells which are sensitive to radiatior and do repair damage. In view of the experimental observations referred to previously, the effect of the sensitizer on the cells in the anoxic compartment will greatly depend upon the particular combination of the radiation dose and concentration of the sensitizing compound. When irradiatior is done with small doses the sensitizer will be protective if present at a low molarity and will have no effect at a high molarity. When irradition is done with large doses the compound will have a sensitizing effect at both low and high molarities. In the hypoxic compartment the cells will be sensitized at all molarities when exposed to either small or large radiation doses. In the oxic compartment the radition response of the cells will not be affected by the sensitizer at any concentration.

These characteristics of the three-compartment model suggest that, if anoxic cells are of importance, the efficient use of the sensitizers requires large radiation doses, and small doses will have little or no effect or possibly even some adverse effects.

Several observations made on in-vivo experiments have been reported which are consistent with this conclusion; however, none of them supports it unequivocally. The survival curve of the KHT tumour cells irradiated after intratumoral administration of NDPP, an efficient sensitizer of hypoxic cells, suggests varying dose-modifying factors at different survival levels [23] in agreement with our conclusion. However, the statistical significance of the variation has yet to be established.

A considerable variation of the dose-modifying factors in relation to the survival levels is indicated also by the survival curves of EMT6 tumour cells determined in experiments in which the host animal was injected with the same sensitizer as used by the present authors [24]. However, it is not possible to distinguish to what extent the results are due to the composite survival curve of a mixed population of oxic and hypoxic cells, or to a difference in the repair capacity of these cells in accordance with our hypothesis. The sensitization of mouse mammary tumours with the same compound as used in our experiments was also found to be less efficient when the tumours were irradiated with fractionated small doses than with single large doses [25]. Reoxygenation during the fractionation interval can be an alternative explanation of this observation. However, the finding that the sensitizer in combination with low dose fractions did not show any significant effect in experiments with poorly reoxygenating osteosarcoma [26], clearly supports our hypothesis. At present it would appear that any definite conclusions on the relevance of the three-compartment model for defining dose fractionation schedules for sensitizer-treated tumours will have to await more quantitative information on the proportion of hypoxic and anoxic cells in tumours, and the time scale of their reoxygenation.

REFERENCES

[1] INTERNATIONAL ATOMIC ENERGY AGENCY, Advances in Chemical Radiosensitization (Proc. Panel Stockholm, 1973), IAEA, Vienna (1974).
[2] ADAMS, G.E., "Electron-affinic sensitizers for hypoxic cells", Proc. XI Int. Cancer Congr. (BUCALOSSI, P., VERONESI, U., CASCINELLI, N., Eds), Excerpta Medica, Amsterdam 5 (1975) 83.
[3] URTASUN, R.C., STURMWIND, J., RABIN, H., BAND, J.R., CHAPMAN, J.D., High dose metronidazole: A preliminary pharmacological study prior to its investigational use in clinical radiotherapy trials, Br. J. Radiol. 47 (1974) 297.
[4] DEUTSCH, G., FOSTER, J.L., McFADZEAN, J.A., PARNELL, M., Human studies with high dose metronidazole: A non-toxic radiosensitizer of hypoxic cells, Br. J. Cancer 31 (1975) 75.
[5] SCOTT, O.C.A., "Radiosensitizers, radiotherapy and fraction size", Fraction-Size in Radiobiology and Radiotherapy (SUGAHARA, T., RÉVÉSZ, L., SCOTT, O.C.A., Eds), Igaku Shoin, Tokyo (1973).
[6] RÉVÉSZ, L., SCOTT, O.C.A., Clinical trials of radiosensitizers and fractionation, Br. J. Radiol. 45 (1972) 626.
[7] DISCHE, S., "Clinical trials with hyperbaric oxygen in radiotherapy", Oncology 1970, Year Book Medical Publishers Inc., Chicago 3 (1972) 308.
[8] PLENK, H.P., "Hyperbaric oxygen radiation therapy. Time-dose schedules and present status", Fraction-Size in Radiobiology and Radiotherapy (SUGAHARA, T., RÉVÉSZ, L., SCOTT, O.C.A., Eds), Igaku Shoin, Tokyo (1973).
[9] HENK, J.M., SMITH, C.W., Unequivocal clinical evidence for the oxygen effect, Br. J. Radiol. 46 (1973) 146.
[10] RÉVÉSZ, L., "Effect of radiosensitizers in relation to the size of radiation dose fraction", Advances in Chemical Radiosensitization (Proc. Panel Stockholm, 1973), IAEA, Vienna (1974) 55.
[11] LITTBRAND, B., Survival characteristics of mammalian cell lines after single or multiple exposures to Roentgen radiation under oxic or anoxic conditions, Acta Radiol., Ther., Phys., Biol. 9 (1970) 257.
[12] LITTBRAND, B., RÉVÉSZ, L., The effect of oxygen on cellular survival and recovery after radiation, Br. J. Radiol. 42 (1969) 914.
[13] RÉVÉSZ, L., LITTBRAND, B., MIDANDER, J., SCOTT, O.C.A., "Oxygen effects in the shoulder region of cell survival curves", Cell Survival at Low Doses of Radiation: Theoretical and Clinical Implications (ALPER, T., Ed.), J. Wiley, Chichester (1975).
[14] RÉVÉSZ, L., LITTBRAND, B., "The composite radioprotective and radiosensitizing effect of an organic free radical", Radiation Protection and Sensitization (MOROSON, H.L., QUINTILIANI, M., Eds), Taylor and Francis Ltd., London (1970) 217.
[15] PETTERSEN, E.O., OFTEBRO, R., BRUSTAD, T., X-ray inactivation of human cells in tissue culture under aerobic and extremely hypoxic conditions in presence and absence of TMPN, Int. J. Radiat. Biol. 26 (1974) 523.
[16] GILLESPIE, C.J., CHAPMAN, J.D., REUVERS, A.P., DUGLE, D.L., "Survival of X-irradiated hamster cells: Analysis in terms of the Chadwick-Leenbouts model", Cell Survival at Low Doses of Radiation: Theoretical and Clinical Implications (ALPER, T., Ed.), J. Wiley, Chichester (1975) 25.
[17] HALL, E.J., ROIZIN-TOWLE, L., Hypoxic sensitizers: Radiobiological studies at the cellular level, Radiology (in press).

[18] EBERT, M., HORNSEY, S., HOWARD, A., Effect of inert gases on oxygen-dependent radiosensitivity, Nature (London) **181** (1958) 613.
[19] LITTBRAND, B., RÉVÉSZ, L., Survival of cells in anoxia, Br. J. Radiol. **41** (1968) 479.
[20] BORN, R., Zellkinetische Untersuchungen an chronisch hypoxischen Fibroblasten des chinesischen Hamsters Gesellschaft für Strahlen- und Umweltforschung mbH, Munich (1974).
[21] SUIT, H., URANO, M., Repair of sublethal radiation injury in hypoxic cells of a C_3H mouse mammary carcinoma, Radiat. Res. **37** (1969) 423.
[22] RÉVÉSZ, L., "Dependence of dose-fractionation on the oxygenation of tumors", Fraction-size in Radiobiology and Radiotherapy (SUGAHARA, T., RÉVÉSZ, L., SCOTT, O.C.A., Eds), Igaku Shoin, Tokyo (1973).
[23] RAUTH, A.M., KAUFMAN, K., In vivo testing of hypoxic radiosensitizers using KHT murine tumor assayed the long-colony technique, Br. J. Radiol. **48** (1975) 209.
[24] BROWN, M.J., Selective radiosensitization of the hypoxic cells of mouse tumors with the nitroimidazoles metronidazole and Ro 7-0582, Radiat. Res. (in press).
[25] ADAMS, G.E., FOWLER, J.F., Nitroimidazoles as hypoxic sensitizers in vitro and in vivo, these Proceedings.
[26] van PUTTEN, L.M., SMINK, T., Effect of Ro-07-0582 and radiation on a poorly reoxygenating mouse osteosarcoma, these Proceedings.

EFFECT OF PEPTICHEMIO ON THE SURVIVAL OF ISOLATED MAMMALIAN CELLS

O. DJORDJEVIĆ, L. KOSTIĆ

Institute of Nuclear Sciences "Boris Kidrič",
Belgrade, Yugoslavia

Abstract

EFFECT OF PEPTICHEMIO ON THE SURVIVAL OF ISOLATED MAMMALIAN CELLS.
Numerous clinical trials have shown that different chemotherapeutic agents have a positive effect in tumour therapy. For these agents to be used effectively, further investigation of the mechanisms of their action at the cytological and molecular level is still needed. One of the new antiblastic compounds, Peptichemio (PTC), which represents a complex of synthetic peptides, has shown high therapeutic effectiveness on various types of malignant tumours. The paper deals with the effect of PTC on the reproductive capacity of mouse L cells studied by the technique of colony-forming units. A typical dose survival curve was obtained at PTC concentrations of 0.1–5 µg/plate. The presented data indicate that the reproductive apparatus of mammalian cells is very sensitive to PTC.

The effect of alkylating agents having chemotherapeutic action is at present the subject of very intensive investigation. It is known that alkylating agents induce many biological effects in living systems very similar to those produced by ionizing or ultraviolet irradiation, such as the killing effect and effects at the macromolecular level on DNA, RNA and proteins [1]. There are data indicating that some bifunctional alkylating agents of a radiomimetic nature can repair sublethal damage induced by these drugs [2]. In radiotherapy as well as in chemotherapy, the repair of sublethal damage may play an important role in the efficiency of fractionated therapy of tumours since the presence of resistant clonogenic cells is one of the serious problems. Only a limited number of chemotherapeutic drugs are efficacious against these cells at concentrations non-toxic to normal growing cells [3]. For this reason the combination of different agents (polychemotherapy), with different modes of action, could also provide better results in the therapy of tumours. Our knowledge of the use of a number of cytostatic drugs in cancer chemotherapy is considerable, but their efficacy will also depend on a better understanding of the mechanisms of their action. The use of isolated mammalian cells in a culture for evaluation of cytotoxicity and biochemical changes could be of value in screening potential antitumour effects in vivo.

This paper deals with the effect of Peptichemio (PTC) on the survival of mammalian cells in vitro. PTC is one of the new chemotherapeutic alkylating agents synthesized by de Barbieri et al. [4]. It is composed of a complex of six peptides of m-dichlorodiethylamino-L-phenyl-alanine (Fig. 1). Numerous clinical trials show that PTC is a very active anti-neoplastic drug. Its positive effects on acute and chronic haemoblastoses as well as on solid tumours have been demonstrated [5]. PTC is characterized by high therapeutic efficacy in treating various types of malignant tumours which predominate in children (neuroblastoma, rhabdomyosarcoma, malignant lymphoma and reticuloendothelioses) [6]. When given alone or in combination with other drugs (5-fluorouracil and actinomycin D), PTC has been used efficiently in the therapy of gynecological cancer [7] and bronchial carcinomas [8].

In contrast to well-documented clinical trials, available data on the mechanism of PTC activity are scarce. Since there are indications that proliferative tumour cells exhibit radiation and drug responses similar to those of cultured cells, we used the system of isolated mammalian cells for studying the effect and mechanism of PTC action.

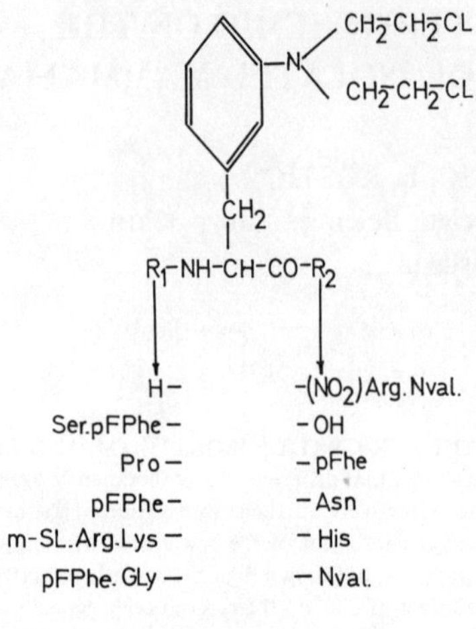

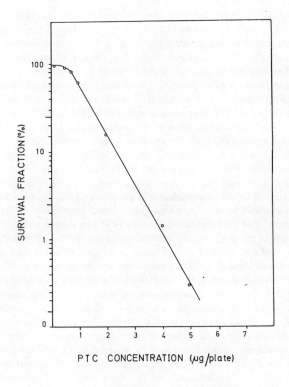

FIG.1. *Chemical formula of Peptichemio (PTC).*

FIG.2. *Dose-survival curve for asynchronous L cells exposed to PTC.*

For this purpose we used a clonal derivative of the established L cell line. Cells were cultivated in modified Eagle's medium, supplemented with 20% calf serum in an atmosphere containing 5% CO_2. The proliferative capacity of the cells treated with different PTC concentrations was studied by their colony-forming ability in vitro. A cell was considered to have retained its reproductive capacity if it gave rise to a colony of 100 cells or more.

The survival of L cells, expressed as the per cent of colonies formed in PTC-treated cultures compared with untreated controls, is presented in Fig. 2.

It is evident that the typical dose survival curve is obtained when cells are treated with PTC in concentrations ranging from 0.1 to 5 μg/plate. Relatively low PTC doses, such as 0.5 μg/plate, reduce the survival of L cells by 10%, whereas that of 0.75 μg/plate reduces the survival by 15%. However, the maximum dose used in these experiments, 5 μg/plate, caused high reduction in survival so that only 0.3% of the cells were able to form colonies. The obtained dose survival curve is similar in form to those produced by X- or u.v.-irradiation.

The results presented here are preliminary. However, further investigation of the mechanism of PTC action at the cytological and biochemical level in mammalian cells in vitro, as well as the joint effect of PTC and irradiation, would provide more information for its efficient application in tumour therapy.

REFERENCES

[1] ROSS, W.C.J., Biological Alkylating Agents, Butterworths, London (1962).
[2] MAURO, F., ELKIND, M.M., Comparison of repair of sublethal damage in cultured Chinese hamster cells exposed to sulfur mustard and X-rays, Cancer Res. 28 (1968) 1156.
[3] SUTHERLAND, R.M., Selective chemotherapy of noncycling cells in an "in vitro" tumor model, Cancer Res. 34 (1974) 3501.
[4] BARBIERI, A.DE., et al., Il Peptichemio: Sintesi di ricerche farmacologiche e biologico-molecolari, in Atti del Simposio sul Peptichemio, Milano, 18 Nov. 1972, I.S.M., Milano (1974) 13.
[5] ASTALDI, G., MEARDI, G., YALCIN, B., KRČ, I., TAVERNA, P.L., KRČOVA, V., Ricerche sperimentali e cliniche con il Peptichemio, ibid. p.75.
[6] MASSIMO, L., PASINO, M., MORI, P.G., COTTAFAVA, F., COMELLI, A., Ruolo del Peptichemio in oncologia pediatrica:ricerche sul comportamento del linfocito, ibid. p.213.
[7] NATALE, N., MANGIONI, C., REMOTTI, G., Prime esperienze nel trattamento dei tumori genitali femminili con Peptichemio, ibid. p.246.
[8] INGRAO, F., CATACCHIO, L., GIORDANO, F., INGIANNA, O., PAOLELLA, E., SPADONI, M., Il Peptichemio, da solo e in combinazione, nella chemioterapia del carcinoma bronchiale, ibid. p.124.

INTERACTION OF LUCANTHONE AND LOW RADIATION DOSES ON MOUSE EMBRYOS IN RELATION TO LET AND DOSE-RATE

Hedi FRITZ-NIGGLI, C. MICHEL
Radiobiological Institute,
University of Zurich,
Switzerland

Abstract

INTERACTION OF LUCANTHONE AND LOW RADIATION DOSES ON MOUSE EMBRYOS IN RELATION TO LET AND DOSE-RATE.

Previous experiments on the protection of radiation- and drug-induced damage of the energy metabolism (liver mitochondria) have shown the importance of energy-dependent repair systems as modifiers of radiosensitivity. Consequently, an analysis was made of the interaction of known oxidative phosphorylation inhibitors and (or) antimetabolites with the very sensitive and probably repair-dependent embryonic system. Bilirubin (hyperbilirubinaemia), iodoacetamide, Reverin and Ledermycin have been found to be radiosensitizing in producing visible malformations in mouse and rat embryos: an impressive synergistic effect of low doses of Lucanthone on 8-day-old mouse embryos was shown. Irradiation with 13.6 rad (high dose-rate, 200 kV) produced anomalies in about 11% of the embryos, 70 mg/kg of Lucanthone alone produced 8.3%, whereas the combination of Lucanthone and 13.6 rad produced 44.5%. The radiation effect itself and the dose enhancement factor depend on LET, as studies with 29-MeV photons, 15-MeV electrons and pions have shown. The negative response of tumour cells agrees with our working hypothesis of the important action of sensitizers and radioprotective substances on immediate and early repair mechanisms depending on the energy metabolism (ATP pool). The possible use of radioprotective substances for healthy tissue and radiosensitizers for tumour tissue is discussed from the standpoint that irradiation alone of mouse embryos with 1 rad could be teratogenic and that irradiation in combination with Lucanthone could be of clinical interest.

INTRODUCTION

The mode of action of radiation protective substances and of radiation sensitizers can vary and can be interpreted in different ways. Therefore, modifications of primary and secondary radiation products — be they water or organic compounds — are possible, as are changes in radiosensitive biomolecules, etc. Previous experiments relating to protective factors and sensitizers of radiation-induced damage of liver mitochondria led us to believe that the energy pool of the irradiated cell and its influence are related to the extent of the radiation damage. An uninjured energy metabolism, characterized among other things by an unhindered ATP (adenosinetriphosphate) production, permits immediate repair processes. Actually, iodoacetamide acts as a radiation sensitizer (even after irradiation) and as an uncoupler of oxidative phosphorylation [1–4] thereby inhibiting ATP production. Similarly, bilirubin — which inhibits, in vitro, the oxidative phosphorylation localized in mitochondria — is probably also a radiation sensitizer in vivo. Gunn rats affected by hereditary hyperbilirubinaemia showed greater damage after irradiation than did normal rats and, among other things, a sensitization of developmental anomalies [5].

This and other considerations led us to examine the radiation-modifying characteristics of other inhibitors of oxidative phosphorylation and of antimetabolites. We assume also that the dependence of the radiation effect on the spatial energy distribution (inexactly described by the linear energy transfer — LET) can be correlated with the influence of the energy metabolism of cells and cell systems. Certainly, radiations do not damage only target biomolecules, but probably

TABLE I. RADIATION-EFFECT CHAIN

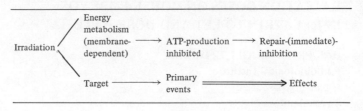

TABLE II. POSSIBLE ACTION OF LET ON ENERGY METABOLISM

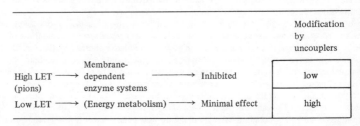

specially the membrane-dependent energy metabolism and the repair systems dependent on the biomolecules (Tables I and II and Fig. 1). For instance, by blocking important repair factors, a high LET radiation could lower the possibility of recovery and thereby act "directly".

The varying LET dependence of different biological reaction systems, or of different phases of the same reaction system, could be interpreted in the sense that there are basically two types of reaction systems [6], namely:

1. LET- and modification-independent systems, characterized by a membrane-independent ATP production and by radiation-insensitive repair systems, respectively.
2. LET- and modification-dependent systems with membrane-dependent repair systems and with radiation-sensitive ATP production, respectively.

Certain types of tumours can belong to the first category and therefore also react differently to radiation sensitizers that are effective in healthy tissue. Therefore, we were interested in the study of an eventual dependence of the radiation-modifying effect of our substances on LET and on the various systems.

In connection with the experiments with iodoacetamide and Gunn rats, the effect of the tetracyclines Reverin and Ledermycin and particularly of Lucanthone was examined. Tetracycline inhibit protein synthesis in microorganisms and mammalian cells and act as uncouplers of oxidativ phosphorylation in mitochondria. They proved to be effective sensitizers of radiation-induced developmental anomalies of the rat [7, 8]. Lucanthone (Miracil D) is an effective drug against schistosomiasis in man and an inhibitor of ribonucleic-acid synthesis in bacteria, fungi and HeLa cells without influencing the protein synthesis [9–12]. It has also proven to be radiation sensitizir for chromosome aberrations as well as chromosome loss in *Drosophila* germ cells and HeLa cells [13–16].

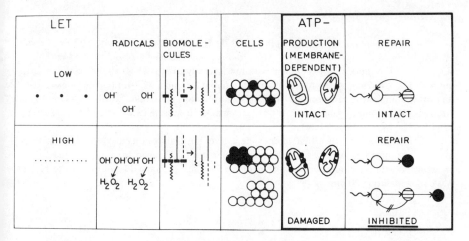

FIG.1. *Possible mechanisms of the different effectiveness of high and low LET radiation. Differences in distribution of primary effects (radicals, biomolecules, cells) and in the damage of the energy and repair metabolisms.*

Besides using cell cultures of the Chinese hamster as test material, we also chose Ehrlich carcinoma cells, embryonic stages of the mouse characterized by their very distinct radiation sensitivity. Moreover, embryonic cells of mammalians seemed to us to belong to the group of reaction systems which is particularly LET- and modification-dependent. Therefore, they are specially suited for analyses that differentiate the LET dependence. As the initial experiments with Lucanthone showed it to have a strong influence on radiation damage, further analysis seemed desirable [17].

METHODS

For the experiments with 9-day-old rat embryos, we chose day 8 of pregnancy as the day for irradiating the Füllinsdorf albino strain of mice. Virgin mice between 2 1/2 and 3 months of age were mated with fertile males of the same strain. The day of finding spermatozoa or a vaginal plug was designated as day 8 of pregnancy. After five days, on day 13 of pregnancy, all the pregnant mice were killed and the foetuses removed for subsequent micro- and macroscopic examination.

Lucanthone, in a dose of 70 mg/kg body weight, was freshly dissolved in sterile water and injected i.m. or i.p. half an hour before irradiation. A lower dosage had previously proved not to be radiation sensitizing. The animals were maintained in a temperature-regulated room at 22°C with a 12-h light-dark cycle and were not anaesthetized for irradiation. Water and food (NAFAG: No. 194) were available ad libitum. There were four types of controls: a control in which the animals were not placed in the irradiation phantom; one in which they were placed in the phantom; one in which the animals were injected with Lucanthone and enclosed in the phantom; and one in which the animals were injected but not enclosed in the phantom. The phantom itself was made of Plexiglas and a Lucite chamber. For the first experiment the Lucite chamber was slightly larger than for the following ones. During irradiation the chambers were ventilated. More details of the experimental procedure can be found in Ref. [17].

IRRADIATION PROCEDURE

The experimental details are given in Ref. [17]. Irradiation was given with 200-kV X-rays, 12 mA, 1 mm Al, 0.5 mm Cu, 37.7 cm target-to-object distance and a dose-rate of 54 rad/min ($\bar{L}_{T100}$ = 1.7 keV/μm). The experiment with low intensity irradiation (0.68 rad/min) was carried out with 140 kV, 6 mA, 1 mm Al, 0.25 mm Cu, 0.5 mm Sn, and a 92.7-cm target-to-object distance. As further comparative radiations, 29-MeV photons (40 rad/min; g-value rad/R = 0.89) and 15-MeV electrons (45 rad/min; g-value rad/R = 0.847) from a betatron were used, which have an extremely low LET ($\bar{L}_{T100}$ = 0.19 keV/μm).

For irradiation with high LET, pions were used in the peak region (star formation) of the 590-MeV proton accelerator of the SIN (Schweiz. Institut für Nuklearforschung). In these first experiments with the biomedical pion beam of the 590-MeV proton a dose of only 12 to 13.6 rad was obtained during 20 min.

RESULTS

Effect of irradiation alone

As can be seen in Figs 2, 3 and 4, when the irradiation is acute, developmental anomalies occur in proportion to the dose for all types of radiation, even with the low doses of 13.6 rad and 1 rad. While the number of resorptions of implanted embryos, i.e. the number of post-implantative deaths, does not increase compared with the control values up to a dose of 50 rad, a clear increase of anomalies was evident even in the lower dose range. The following anomalies were observed in the surviving 13-day-old foetuses: growth reduction, disturbances in eye formation, i.e. unilateral or bilateral microphthalmia and anophthalmia, as well as microcephaly (Table III).

It is particularly worth noting that even a dose of 1 rad — in this case protracted and applied during 1 min and 28 s — produced up to 18.1% anomalies (mainly microphthalmia) [17]. The 200-kV X-rays are unmistakably more effective than the betatron rays which showed interesting differences between the effects of the electrons and photons. The effect, per rad, of the electrons is greater than that of the 29-MeV photons. Although, according to physical considerations, the LET of the 15-MeV electrons hardly differs from that of the 29-MeV photons, it seems that the embryonic cells can distinguish between these two types of radiation and presumably record imperceptible differences in the spatial energy distribution. We have already observed these phenomena several times and intend to study them in more detail. At a dose of 100 rad, the effect both of the electrons and of the photons of the betatron is clearly inferior to that of the 200-kV X-rays. Also, the dose-dependence of the RBE factor has already been frequently observed by us.

Pion irradiation was only possible at a low intensity. In the peak region the pion irradiation represents a high LET radiation which is composed of neutrons, alpha rays and protons of disintegrated atomic nuclei. For comparison, 140-kV X-rays of equal intensity were used. The pion irradiation (Fig. 3), with an RBE factor of about 2, was more effective than the 140-kV X-rays. At the same time, an interesting increase in effectiveness of the protracted irradiation was noticed compared with the acute irradiation. However, it appeared that this difference is not a real one because the 30-min confinement of the mice in the ventilated but narrow irradiation chamber led to a considerable increase of damage even without irradiation. This "cage effect" was observed by us for the first time and will be examined further. We assume that the stress factors act as sensitizers and that these factors may be of endocrine origin. If we take into account this cage effect, the effect of the protracted irradiation becomes practically negligible.

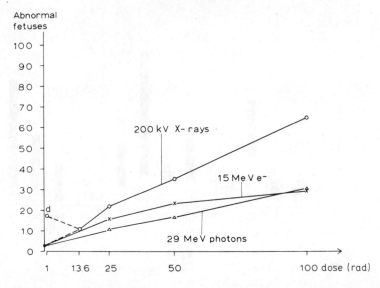

FIG.2. Incidence of anomalies in mouse 13-d foetuses irradiated as 8-d embryos with 200-kV X-rays, 29-MeV photons or 15-MeV electrons (40-54 rad/min); d = irradiation with 0.68 rad/min.

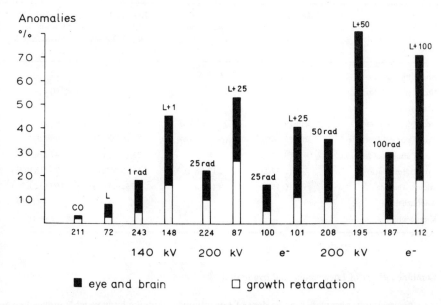

FIG.3. Enhancement of the incidence of anomalies by treatment with Lucanthone (70 mg/kg body weight) and 12–13.6 rad. L + I = Application of Lucanthone ½ h before irradiation. I + L = Application of Lucanthone immediately after irradiation (200 kV = 54 rad/min, 140 kV = 0.68 rad/min, π^- = pi-mesons = 0.68 rad/min).

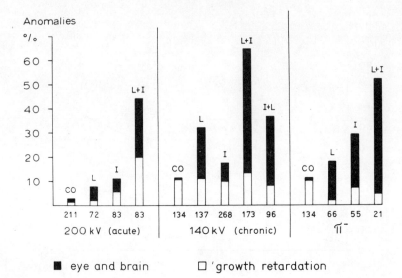

FIG.4. *Enhancement of the incidence of anomalies by Lucanthone (70 mg/kg body weight) given $\frac{1}{2}$ h before irradiation with 140-kV X-rays, 200-kV X-rays and 15-MeV electrons.*

TABLE III. SENSITIZATION OF THE RADIATION EFFECT (50 rad, 200 kV, 54 rad/min) BY LUCANTHONE

	Untreated	Treated with Lucanthone	Irradiated with 50-rad, 200-kV X-rays	Treated with Lucanthone and 50-rad, 200-kV X-rays
No. of mothers	17	6	15	15
No. of implantations	222	76	223	210
Resorptions	11 (4.95%)	4 (5.26%)	15 (6.73%)	15 (7.14%)
No. of foetuses tested	211 (95.05%)	72 (94.74%)	208 (93.27%)	195 (92.86%)
No. of normal foetuses	204 (96.68%)	66 (91.66%)	135 (64.90%)	40 (20.52%)
No. of abnormal foetuses	7 (3.32%)	6 (8.34%)	73 (35.10%)	147 (79.48%)
Growth retardation	5 (2.37%)	2 (2.78%)	18 (8.65%)	36 (18.46%)
Microphthalmia	2 (0.45%)	4 (5.56%)	55 (26.45%)	103 (52.82%)
Anophthalmia	–	–	–	8 (4.10%)
Microcephaly	–	–	–	8 (4.10%)

Combined effect of Lucanthone and irradiation

As previously observed, a single dose of Lucanthone (70 mg/kg body weight) induced several anomalies. Like tetracyclines, Lucanthone is teratogenous. Microphthalmia prevail over growth reductions (see Table III and Figs 3, 4).

TABLE IV. SENSITIZATION EFFECTS OF LUCANTHONE

Treatment		Anomalies (%)
1 rad (chronic)	140 kV	42
13.6 rad (acute)	200 kV	57
13.6 rad (chronic)	140 kV	24
13.6 rad (chronic)	π^-	10
25 rad (acute)	200 kV	42
25 rad (acute)	15 MeVe$^-$	40
50 rad (acute)	200 kV	45
100 rad (acute)	15 MeVe$^-$	46

Again, the observation that there is also a cage-effect with a single dose of Lucanthone is of exceptional interest. Animals that had been enclosed in a phantom for 30 min showed a clear increase in anomalies, in particular up to 18.3% in the pion experiments and up to 32.1% in the experiments which were carried out during the autumn with narrower chambers. This phenomenon is being further thoroughly examined by us. Qualitatively, the teratogenic effects of Lucanthone alone and of radiation alone do not differ.

When Lucanthone was administered 1/2 hour before irradiation, the irradiation effect was enhanced for all radiation doses. The effect was not additive but synergistic. For all doses and types of radiation examined, a sensitizing effect of Lucanthone was observed, though the degree of enhancement differed (Table IV). The sensitizing factor as a percentage of the enhancement of the total effect was calculated according to the following formula: effect (L + irradiation) - (L effect alone + irradiation effect alone) × 100 : effect (L + irradiation).

After subtracting the so-called cage-effect from the apparent effect of the chronic doses, the lowest enhancing effect was produced by pion irradiation and the highest by concentrated 200-kV X-ray irradiation with 13.6 rad.

The enhancing effect with electrons was somewhat less. Interestingly, the effect of a combined Lucanthone and irradiation treatment differs qualitatively from the treatments alone. In some cases, exencephalies appeared and, besides a change in the ratio of microphthalmias and growth reductions, there were changes to the advantage of the eye anomalies. Thus, the combined treatment produced mainly eye anomalies and the types of brain malformations mentioned.

It is interesting to note that Lucanthone can even sensitize the effect of 1 rad (difference highly significant in χ^2 test).

Comparison of the sensitizing effects of bilirubin, iodoacetamide and tetracyclines

In sensitizing radiation-induced embryonic damage, Lucanthone proved to be the most effective substance compared with animals with hyperbilirubinaemia or after administration of iodoacetamide, Reverin or Ledermycin. The effects of combined therapy with these agents were not qualitatively different from those observed for the treatments with the agents alone. Naturally, we realize that the comparisons could only be made summarily because all the dose-effect curves for the various agents could not be completed by us as the experiments are extremely time-consuming.

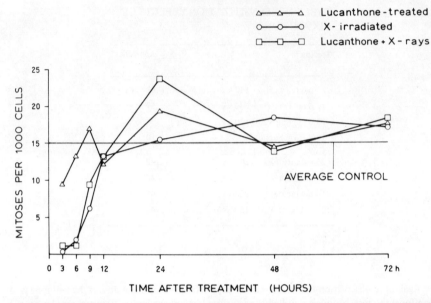

FIG.5. Effect of treatment with Lucanthone (70 mg/kg body weight) and X-rays (300 rad) on mitosis of Ehrlich ascites tumour cells [18].

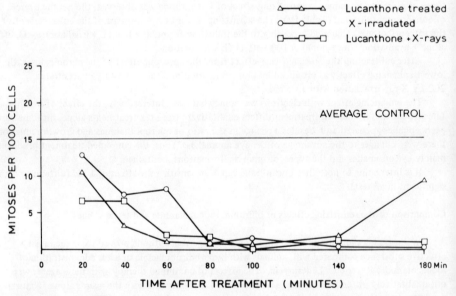

FIG.6. Effect of treatment with Lucanthone (70 mg/kg body weight) and X-rays (300 rad) on mitosis of Ehrlich ascites tumour cells [18].

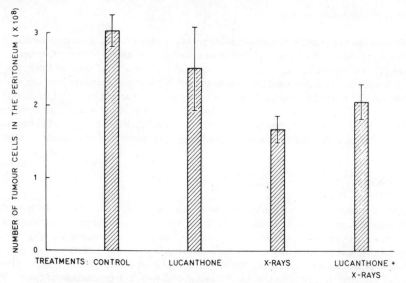

FIG.7. Growth-response of Ehrlich ascites carcinoma treated with Lucanthone and X-rays (750 rad) [18].

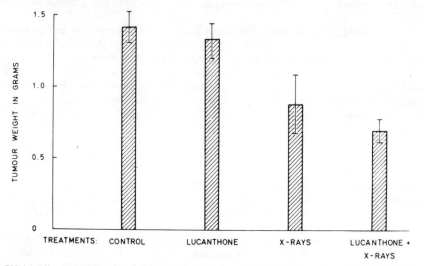

FIG.8. Effect of Lucanthone and X-rays (300 rad) on the growth of Ehrlich carcinoma in solid form [18].

Effect of a single dose of Lucanthone after irradiation

As, according to our working hypothesis, radiation sensitizers intervene in the repair processes by affecting the energy metabolism, we were interested in knowing whether also Lucanthone acts as sensitizer when administered immediately after irradiation. A tentative experiment with 13.6 rad of protracted irradiation actually showed that a single dose of Lucanthone given after irradiation had a sensitizing effect (Fig. 4).

Effects on other reaction systems

Reverin did not sensitize radiation-induced growth retardation of Ehrlich carcinoma in the solid form, nor the mitotic activity of Ehrlich carcinoma cells in ascites form nor Crocker sarcoma [7]. On the other hand, interesting differences in the sensitizing effect of Reverin could be seen in relation to chromosomal aberrations and cell death of hamster cells. Four-day-old in-vivo cell cultures of Ehrlich carcinoma cells were not sensitized by pretreatment with Lucanthone (see Figs 5, 6 and 7). There was no difference between the mean weight of solid tumours after irradiation and after combined treatment (Fig. 8). Also, the irradiation damage in bilirubin-treated animals compared with normal rats was enhanced differently according to the reaction system [5].

DISCUSSION

Our experiments showed a strongly enhancing effect of Lucanthone on the radiation-induced developmental damage in mice. Lucanthone proved to be a stronger sensitizer than iodoacetamide and tetracyclines. The effect of Lucanthone after irradiation also supports our working hypothesis that certain radiation sensitizers affect the immediate repair processes. The strong modifying power of the investigated agents can be explained by the particular dependence of embryonic systems on the repair capacity. It is possible that Lucanthone also inhibits ATP production indirectly. We are investigating this problem at the present time. The sensitizing effect of Lucanthone even with the small dose of one rad is particularly interesting, specially in connection with problems of radiation protection.

It is interesting that actinomycin, which resembles Lucanthone structurally and biochemically [19], increased radiation-induced injury in normal tissues of mice [20]. Moreover, the extent of the sensitization through combined treatment is dependent on the spatial energy distribution, i.e. on LET. A high LET irradiation (pions in the peak region) can indeed — according to our tentative experiments — be modified, but only to a slight extent. The modification possibility is particularly interesting in relation to the planned radiation therapy with pions. We assume that with a concentrated irradiation, sensitization is eventually augmented. The LET dependence of the

TABLE V. EFFECTIVE MODIFIERS FOR RADIATION THERAPY

	Objective	Characteristics of modifiers
Radioprotective substances for healthy tissues	Protection of membrane-dependent energy metabolism	Membrane-protectors, Protectors of mitochondrial functions (e.g. rutosides, EDTA etc.)
Radiosensitizers for tumour tissues	Inhibition of membrane-independent repair systems	Inhibitors of glycolysis
Tumour tissue with high oxidative phosphorylation	Inhibition of membrane-dependent repair systems	*Uncoupler* Selective action on tumour tissue required

sensitizing effect supports our working hypothesis, according to which a high LET irradiation additionally disturbs the energy systems. The already strongly damaged systems cannot be damaged further by other agents.

The observations that embryonic cells can be sensitized, whereas the tumour cells examined by us cannot, also support our hypothesis. It is possible that tumour cells are more independent of the inhibitions of the membrane-dependent ATP production (oxidative phosphorylation) as they can draw their energy from glycolytic processes as well. Naturally, it is possible that certain other types of tumours could be sensitized as in a clinical instance where an adjuvant effect of Lucanthone on radiation therapy was observed [21].

In the opinion of the authors, an effective modification of radiation therapy of tumours could be achieved if healthy cells, with their membrane-dependent energy metabolism, could be safeguarded by protective substances (Table V). Such substances could eventually be represented by rutosides which protect mitochondria [22]. For certain tumour cells it would be necessary to find sensitizers also capable of damaging the membrane-independent repair systems.

The cage effect showed that even indirect actions can influence pharmacological and radiation effects (see also Ref. [23]) and must be seriously considered in every experiment with radiomodifiers.

ACKNOWLEDGEMENTS

The authors are grateful to Dipl.Phys. Imogen Riehle and Dr. H. Blattmann for their help and to Miss E. Kundert for technical assistance. They would also like to acknowledge the financial support received from the Swiss National Foundation for Scientific Research.

REFERENCES

[1] FRITZ-NIGGLI, Hedi, Post-irradiation induced sensitization of inhibition of oxidative phosphorylation by iodoacetamide, Experientia 22 (1966) 666.
[2] FRITZ-NIGGLI, Hedi, MICHEL, C., "Chemical sensitization of the damaging effects on embryos produced by low radiation doses: the role of energy metabolism and immediate repair", Radiation Protection and Sensitization (PAOLETTI, R., VERTUA, R., Eds), Taylor & Francis Ltd., London (1970) 311.
[3] MICHEL, C., Chemische Sensibilisierung von Strahlenschädigungen bei Rattenembryonen, Archiv der Julius Klaus-Stiftung für Vererbungsforschung, Sozialanthropologie und Rassenhygiene 43/44 (1969) 3.
[4] QUINTILIANI, M., BOCCACCI, M., Effetti del trattamento con gli acidi bromo e iodoacetico sulla mortalità da irradiazione roentgen nel topo, Rend. Ist. Sup. Sanità 23 (1960) 5.
[5] FRITZ-NIGGLI, Hedi, MICHEL, C., NECK, K., STAHEL, A., VOLLENWEIDER-WEPFER, Regular hereditary metabolic anomaly (Hyperbilirubinaemia) as radiosensitizer, Radiat. Environ. Biophys. 11 (1974) 195.
[6] FRITZ-NIGGLI, Hedi, Die Bedeutung des Repairsystems für die relative biologische Wirksamkeit von Strahlen verschiedener Qualität; eine Zwei-System-Theorie, Strahlentherapie 135 (1968) 202.
[7] FRITZ-NIGGLI, Hedi, MICHEL, C., Sensibilisierung der Strahlenschädigung von Rattenembryonen durch Dimethylchlortetracyclin (Ledermycin), Atomkernenergie 18 (1971) 105.
[8] FRITZ-NIGGLI, Hedi, MICHEL, C., RAO, K.R., Sensitizing effect of Reverin (Pyrrolidinomethyltetracycline) on the radiation damage of rat embryos as compared to its effect on other cell systems, Agents and Actions 4 1 (1974) 54.
[9] BASES, R., MENDEZ, F., Reversible inhibition of ribosomal RNA synthesis in HeLa by Lucanthone (Miracil D) with continued synthesis of DNA-like RNA, J. Cell. Physiol. 74 (1969) 283.
[10] WEINSTEIN, B., CHERNOFF, R., FINKELSTEIN, I., HIRSCHBERG, E., Miracil D: an inhibitor of ribonucleic acid synthesis in Bacillus subtilis, Mol. Pharmacol. 1 (1965) 297.
[11] WEINSTEIN, B., CARCHMAN, R., MARNER, E., HIRSCHBERG, R., Miracil D: effects on nucleic acid synthesis, protein synthesis, and enzyme induction in Escherichia coli, Biochim. Biophys. Acta 142 (1967) 440.
[12] HIRSCHBERG, E., CECCARINI, M., OSNOS, M., CARCHMAN, R., Effects of inhibitors of nucleic acid and protein synthesis on growth and aggregation of the cellular slime mold Dictyostelium discoideum, Proc. Natl. Acad. Sci. USA 61 (1968) 316.

[13] BASES, R., Enhancement of X-ray damage in HeLa cells by exposure to Lucanthone (Miracil D) following radiation, Cancer Res. **30** (1970) 2007.
[14] LÜERS, H., Biologisch-genetische Untersuchungen über die Wirkung eines Thioxanthonderivates an *Drosoph melanogaster*, Z. Vererb. Lehre **87** (1955) 93.
[15] U, R., Action of Miracil D (1-diethylaminoethylamino-4-methyl-10-thioxanthenone) and chromosome dama in *Drosophila* male germ cells (Abstract), Proc. 12th Int. Congr. Genetics, Tokyo **1** (1968) 88.
[16] U, R., Inhibitor of ribonucleic acid synthesis and chromosome loss in *Drosophila* male germ cells, Drosophil Inform. Serv. **45** (1970) 161.
[17] MICHEL, C., Combined effects of Miracil-D and radiation on mouse embryos, Experientia **30** (1974) 1195.
[18] RAO, K.R., FRITZ-NIGGLI, Hedi, Effect of Lucanthone on the radiosensitivity of a mouse tumour, to be published in Br. J. Cancer.
[19] EPIFANOVA, Olga I., MAKAROVA, G.F., ABULADZE, M.K., A comparative study of the effects of Lucan (Miracil D) and actinomycin D on the Chinese hamster cells grown in cultures, J. Cell. Physiol. **86** (1975) 26
[20] PHILLIPS, T.L., WHARAM, M.D., MARGOLIS, L.W., Modification of radiation injury to normal tissues by chemotherapeutic agents, Cancer **35** (1975) 1678.
[21] TURNER, Sophie, BASES, R., PEARLMAN, A., NOBLER, M., KABAKOW, B., The adjuvant effect of Lucanthone (Miracil D) in clinical radiation therapy, Radiology **114** (1975) 729.
[22] FRITZ-NIGGLI, Hedi, Schutzwirkung von 0-(β-hydroxyäthyl)-Rutosid gegen die strahleninduzierte Hemm des Energiestoffwechsels, Praxis **57** (1968) 180.
[23] VALENTINI, E.J., HAHN, E.E., The indirect effect of radiation on embryonic mortality, Int. J. Radiat. Bio (1971) 259.

EFFECT OF Ro 07-0582 AND RADIATION ON A POORLY REOXYGENATING MOUSE OSTEOSARCOMA*

L.M. van PUTTEN, T. SMINK
Radiobiological Institute TNO,
Rijswijk,
The Netherlands

Abstract

EFFECT OF Ro 07-0582 AND RADIATION ON A POORLY REOXYGENATING MOUSE OSTEOSARCOMA.
The effect of the application of the hypoxic sensitizer Ro 07-0582 was studied during fractionated experimental radiotherapy of a poorly reoxygenating mouse osteosarcoma. Although the studies are incomplete, the following tentative conclusion can be drawn from the preliminary studies: (1) Application of the drug by i.p. injection 1 hour before radiation exposure seems to enhance growth delay after single dose treatment with 1000 rad and after five fractions of 300 rad per week given in one week. (2) No enhancement of growth delay is observed after two or more weeks of multifraction radiotherapy with 300 rad per fraction, but growth delay is significant if six fractions of 800 rad are applied in two doses per week. It is not clear whether the latter effect is a consequence of the higher doses of sensitizer tolerated when few fractions per week are given or, alternatively, a consequence of the higher radiation dose per fraction. (3) The studies suggest that the conventional five-fractions-per-week schedule may not be optimal for combination with hypoxic sensitizers, but additional studies are necessary to confirm this.

INTRODUCTION

The nitroimidazoles Metronidazole and the compound Ro 07-0582 have been shown to be effective in sensitizing hypoxic cells in vitro and in vivo. Most studies have been performed using single doses of irradiation but recently also the effect of fractionated treatment has been studied by Sheldon et al. [1] for three fractions in 4 days and five fractions in 9 days. Application of these agents in combination with irradiation treatments of 200 to 300 rad per fraction was studied by Brown [2] but not found of benefit, presumably due to extensive reoxygenation of the mammary tumour studied. On the other hand, it cannot be excluded that among the "radioresistant" tumours treated in patients, a subgroup exists in which the reoxygenation phenomenon is less effective or slower than in the majority of tumours, just as has been described for the poorly reoxygenating mouse osteosarcoma C22LR. For this reason, a study of the effect of sensitizers on the results of radiotherapy applied in small fractions to this tumour appeared to be of interest.

This paper is an interim report on studies in progress; experiments have not been completed but the results obtained so far permit the formulation of a number of tentative conclusions.

---

* This work was performed under Contract number NO 1-CM-53763 with the National Cancer Institute, Department of Health, Education and Welfare.

MATERIALS AND METHODS

Osteosarcoma C22LR has been described earlier [3–4]; the tumour is non-antigenic in its host of origin, a (C57BL/Rij × CBA/Rij)F1 hybrid mouse. The tumour has been stored in liquid nitrogen, from which a new sample of passage 76 or 77 has been thawed every two or three months for further studies to ensure the stability of the tumour material over a 10-year period. The thawed suspension is inoculated subcutaneously and passages are transferred four to six times before the line is discarded. Experiments were performed in male 10- to 20-week-old (C57BL/Rij × CBA/Rij)F1 hybrid mice, usually six or seven days after subcutaneous inoculation of one million tumour cells obtained by the cell dispersion technique of Reinhold [5]. The location of the tumour inoculum was on the dorsal flank so as to permit radiation exposure of the full tumour area in unrestricted anaesthetized mice without exposing radiosensitive normal tissues except skin. Irradiation was initially performed on mice anaesthetized with Hypnorm®, 2.5 ml/kg i.p., and later with Nembutal® 1 ml/kg i.p. The former solution contains per ml 10 mg fluanison and 0.2 mg fentanyl base; the latter contains per ml 60 mg of pentobarbital sodium. The exposure conditions were 300 kVp, 10 mA, HVL of the beam 3 mm Cu, dose-rate 350 rad/min. The shielded major part of the body of the mice received less than 2% of the dose given to the tumour. Tumours were measured at least twice a week by determining the diameter with calipers in three dimensions. Each treatment group consisted of five to ten mice. The product of the three diameters was used as an indication of relative tumour volume by comparing it with the mean product at the time at which treatment started. Measurements were stopped when the tumours ulcerated.

Drugs. Ro 07-0582 was administered i.p. about one hour before the start of irradiation. In each experiment a few of the mice died from the combined exposure to Ro 07-0582 and anaesthesia. Although single doses of 1.2 g/kg of the compound Ro 07-0582 were well tolerated, daily administration of this drug dose proved too toxic and doses were reduced accordingly.

RESULTS AND DISCUSSION

The effect of a single dose of 1000 rad X-rays with and without the administration of 1200 mg/kg body weight of Ro 07-0582 is shown in Fig.1. The chemical alone does not cause a difference in tumour growth from the control. The radiation alone causes a delay of growth to twice the original volume of about 5 days ($7\frac{1}{2}$ days instead of $2\frac{1}{2}$ days for the non-treated control) and the delay after combined treatment is more than twice as long. The exact delay could not be measured due to the unexpected skin ulceration over the tumours observed on day 17. A similar delay after fractionated treatment is seen from the data in Fig.2. In this experiment 5 doses of 300 rad were combined with daily treatment with Ro 07-0582 in doses which caused no increased mortality when combined daily with the neuroleptanalgesia (Hypnorm) applied to permit immobilization of the mice during tumour irradiation. Note that these doses were much lower than in the preceding experiment; as also noted by others the sensitizer enhances anaesthesia and causes death in anaesthesia. The growth delay to double the tumour volume after fractionated irradiation alone is 3 days; from 2 days after start of the treatment as observed in the control groups to 5 days in the irradiated group. Again ulceration prevented an accurate measurement of the effect of combined treatment but the delay is more than 8 days in that group.

Problems in interpretation of our results became manifest when we gave the sensitizer in combination with 3- or 4-week treatment schedules as shown in Figs 3 and 4. Three seemingly reproducible effects were noted here: (1) small tumours ulcerate and make the evaluation of regrowth impossible; (2) an initial seemingly more pronounced response of combination treatment than of radiation alone seems to be abolished at the end of the second treatment week; (3) there seems to be a consistent effect of the sensitizer alone; this effect was already suggestive after a single week of treatment (Fig.2), but in that case it seemed to be reversible.

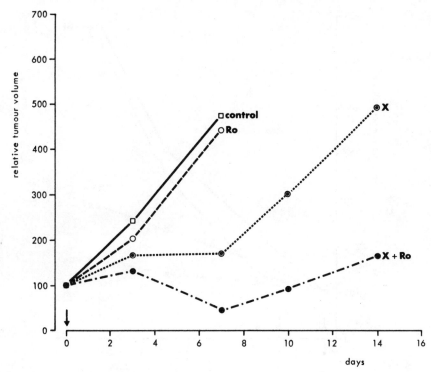

FIG.1. *Tumour growth curve for different groups of mice carrying osteosarcoma implants in the flank and exposed as indicated on the figure.*
Ro = Ro 07-0582 only. X = irradiation only.
X + Ro = combined treatment.
Anaesthesia with Hypnorm.

To prevent ulceration of small tumours an attempt was made to obtain growth delay data in more detail by reducing the radiation dose per fraction to 200 rad (Fig.5). This caused too rapid growth and after the completion of two weeks of treatment accurate measurement of the large ulcerating tumours was no longer possible. These results were in marked contrast with measurements made in 1967 (Fig.6) performed on the same tumour line. In the earlier studies no ulceration and no reduction of tumour volume below that at the start of treatment had been observed, in contrast to the recent results. A change in the tumour seemed unlikely since the cell inoculum in 1967, as in 1975, had been derived from a similar sample from a large batch of tumour cell suspension ampoules stored in liquid nitrogen over the years. Further comparison of the experimental techniques over the years pointed to the possibility that differences in anaesthesia could be responsible for the different effects. Earlier [4] it had been observed that the Nembutal

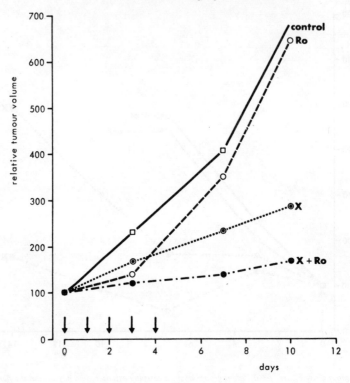

FIG.2. *Tumour growth curves as in Fig.1.*

anaesthesia applied in 1967 caused a decreased sensitivity as measured in an osteosarcoma cell-survival study after fractionated irradiation, and this decreased sensitivity seemed most likely due to increased hypoxia or less reoxygenation of hypoxic cells. On the basis of these findings the possibility was considered that Nembutal anaesthesia might also decrease the sensitivity of the skin by causing hypoxia and thus prevent early ulceration. For this reason Nembutal anaesthesia was substituted for the neuroleptanalgesic mixture and with this regimen a two-week treatment was performed as shown in Fig.7. This shows: (1) early a more pronounced response of the combined treatment group, but the tumour regrowth curve after ending treatment strongly suggests that this is only a temporary effect; (2) again, Ro 07-0582 given alone seems to have an effect.

At this point in our studies we found it very hard to explain the results. For the nitroimidazo a direct cell killing had been observed on hypoxic cells [6], but the absence of this effect in those studies in which volume measurements could be performed after the end of drug exposure

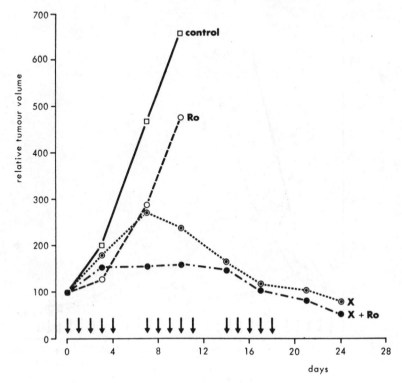

FIG.3. *Tumour growth curves as in Fig.1.*

(Figs 1 and 2) made us doubt the validity of this interpretation. The alternative was the possibility that the drug gave a temporary decrease of tumour volume without really affecting cell survival, as was for instance shown for agents such as vitamin A which cause no change in cell survival but only a more rapid clearance of dead and dying cells probably through stimulation of macrophage activity. This could also explain the early difference between radiation alone and combined treatment in Fig.7 by assuming that the sensitizer in this study was not effective in reducing cell survival. However, this seemed incompatible with the conclusion from the experiments depicted in Figs 1 and 2, which suggested a real effect of the sensitizer, not only after a single high dose, but also after fractionated treatment. Various possibilities suggested themselves to explain the possibility of a real but temporary effect of the compound.

In the first place the possibility was considered that the compound, although radiosensitizing for the hypoxic cells, might at the same time be radioprotective to well-oxygenated cells. A statistically non-significant protection of well-oxygenated mouse spleen colony-forming

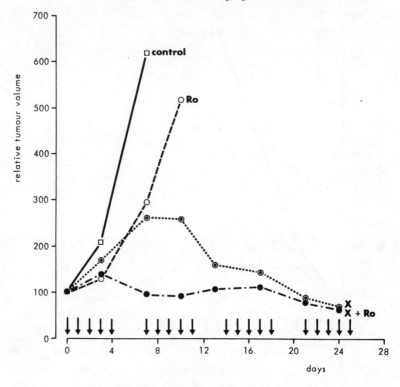

FIG. 4. *Tumour growth curves as in Fig. 1.*

haemopoietic stem cells was observed by Keizer (unpublished). If, after an initial period of treatment, cell survival is markedly reduced, the fraction of hypoxic clonogenic cells might be reduced to make the protective effect dominant in the second half of the treatment period and thus neutralize the sensitization occurring in the initial phase of treatment. No solid arguments for this possible explanation have been brought forward at present and it seems difficult to test it.

In the second place it has been reported that the nitroimidazoles are affected by the drug-metabolizing enzyme system in the liver and it could be postulated that this would cause an increasingly rapid breakdown of active drug; possibly even enhanced by the activation of this enzyme system by the simultaneous application of a barbiturate. To test for this possibility the effect was studied of the drug SKF 525A, a blocker of microsomal drug-metabolizing enzymes. Figure 8 shows that this drug did not modify the final effect of the sensitizer. Because of a

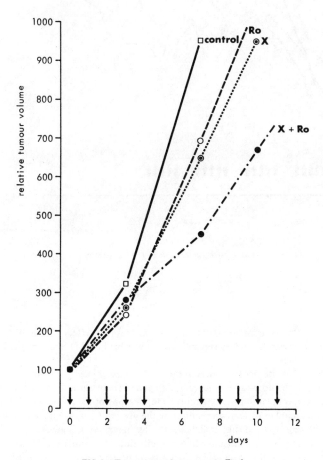

FIG.5. *Tumour growth curves as in Fig.1.*

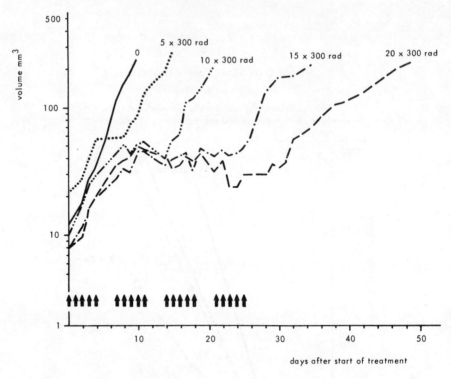

FIG.6. *Tumour growth curves made in 1967. (For explanation see text.)*

technical error in this experiment no data were available for the effect of radiation alone, but the early difference between the curves suggests that a probable initial effect of the sensitizer may indeed have been increased.

In the third place Révész [7] observed that TAN, another electron-affinic sensitizer, not only affects the slope of the dose-survival curve by an apparent substitution for oxygen, as expressed in an enhancement of cellular sensitivity expressed in the slope of the curve, but also seems to modify the extrapolation number of the curve indicating the possibility of an apparent substitution for oxygen in facilitating repair of sublethal damage. This observation, if generally valid for hypoxic sensitizers, could explain the ambiguity in our results since the former would explain sensitization of the hypoxic cells, and the latter affect their relative protection by facilitating repair. It is obvious that the facilitation of repair should be less marked if fewer large radiation doses are given instead of many small doses. A favourable effect of Ro 07-0582 after few large fractions has already been observed by Sheldon et al. [1]. A similar study was performed with our osteosarcoma by applying the sensitizer in combination with six fractions of 800 rad given twice weekly; another group was studied in which the sensitizer was given only during the first three of the six radiation treatments. The results, shown in Fig.9, indicate a significant effect of this type of treatment. Of course this favourable effect may be explained in part by the fact that the more widely spaced treatments permit the use of a higher drug dose per treatment than in combination with the daily treatment and further studies are obviously needed to confirm and

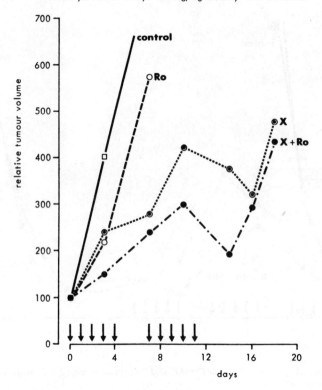

FIG. 7. *Tumour growth curves as in Fig. 1, but anaesthesia with Nembutal.*

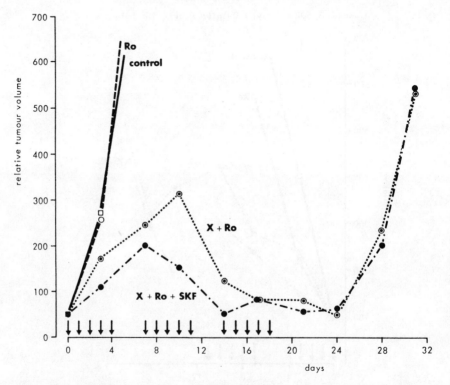

FIG. 8. *Tumour growth curves as in Fig. 7. SKF = SKF 525a, 2-diethylaminoethyl-2, 2-diphenyl valeriate given in a dosage of 32 mg/kg twice weekly.*

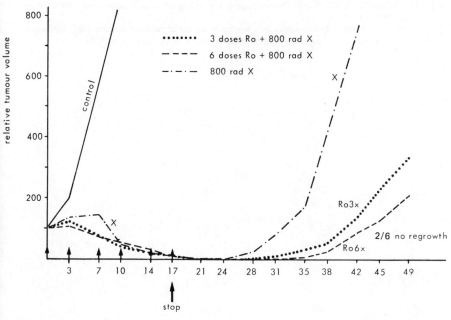

FIG.9. Tumour growth curves as in Fig. 7. Ro 3x and Ro 6x means that Ro 07-0582 was given i.p. 1 hour before the first three of six and before all of six radiation treatments, respectively. 2/6 no regrowth means that in two of six mice in this group no palpable tumour was observed up to day 50 after start of treatment.

expand these data. Nevertheless, these preliminary results suggest that indeed the combination of tolerated doses of Ro 07-0582 with radiotherapy in traditional five-treatments-per-week schedules may be less effective than the combination of the sensitizer with fewer, more widely spaced, fractions of treatment.

ACKNOWLEDGEMENTS

The capable assistance of Metha Pijpers, Julia van Ettinger, Els Over and Leonie van der Burg is gratefully acknowledged. A gift of compound Ro 07-0582 from Messrs Hoffman-Laroche, Basel made these studies possible.

REFERENCES

[1] SHELDON, P.W., FOSTER, J.L., FOWLER, J.F., Radiosensitisation of C3H mouse mammary tumours using fractionated doses of X-rays with the drug Ro-07-0582, Br. J. Radiol. (to be published).
[2] BROWN, J.M., Effect of hypoxic sensitizers, metronidazole and Ro-07-0582 on the radiation response of normal and malignant cells in the mouse. (Abstract), Radiat. Res. 62 (1975) 559.
[3] PUTTEN, L.M., van, Tumour reoxygenation during fractionated radiotherapy; studies with a transplantable mouse osteosarcoma, Eur. J. Cancer 4 (1968) 173.

[4] PUTTEN, L.M., van, LELIEVELD, P., BROERSE, J.J., Response of a poorly reoxygenating mouse osteosarcoma to X-rays and fast neutrons, Eur. J. Cancer 7 (1971) 153.
[5] REINHOLD, H.S., A cell dispersion technique for use in quantitative transplantation studies with solid tumours, Eur. J. Cancer 1 (1965) 67.
[6] SUTHERLAND, R.M., Selective chemotherapy of noncycling cells in an in vitro tumor model, Cancer Res. 34 (1974) 3501.
[7] RÉVÉSZ, L., "Effect of radiosensitizers in relation to the size of radiation dose fraction", Advances in Chemical Radiosensitization (Proc. Panel Stockholm, 1973), IAEA, Vienna (1974) 55.

CLINICAL TRIALS OF Ro 07-0582 AS A SENSITIZER OF HYPOXIC CELLS

I. LENOX-SMITH
Clinical Research Department,
Roche Products Ltd,
Welwyn Garden City, Herts,

S. DISCHE
Regional Radiotherapy Centre,
Mount Vernon Hospital,
Northwood, Middx,
United Kingdom

Abstract

CLINICAL TRIALS OF Ro 07-0582 AS A SENSITIZER OF HYPOXIC CELLS.
    Ro 07-0582 (1-(2-nitro-1-imidazolyl)-3-methoxy-2-propanol) is a 2-nitroimidazole derivative. Volunteer studies in man showed that single oral doses of 1–4 g were rapidly absorbed: there was a linear relationship between the drug serum concentration at 4 h and the dose administered expressed in mg/kg. The serum half-life (mean 12.1 ± 2 h) was independent of dose. The serum concentrations at 2–4 h suggested that a dose of 78 mg/kg would give a serum concentration of 100 μg/ml at 4 h, this being the minimum predicted level at which radiosensitization would be observed. Eight patients were treated with single oral doses of 4–10 g (81–165 mg/kg) and serum levels in excess of 100 μg/ml were obtained in all cases. Immediate gastro-intestinal side-effects appeared to limit the dose to below 140 mg/kg, a dose giving serum levels of approximately 200 μg/ml. Tumour concentrations, determined in five cases, ranged from 12–92% of the serum concentration. In six patients the radiosensitizing properties of Ro 07-0582 were tested by observing the skin reaction after irradiation using a radiostrontium source with and without administration of the drug under both hypoxic and oxic conditions. With doses of 81–165 mg/kg the relative sensitizing efficiency ranged from 27–71%: studies with metronidazole gave figures of 11–15%. Tumour response was measured in seven patients with metastatic disease. Quantitative evidence in two patients showed an enhancement of 20% in radiation response in a patient with multiple skin nodules but no enhancement in one with pulmonary metastases. Qualitative evidence from two other patients suggested some enhancement of radiation effect while the remaining three patients died before any conclusion could be drawn. These results suggest that Ro 07-0582 may prove to be a radiosensitizer of value in clinical radiotherapy and further studies using multiple doses have been initiated.

INTRODUCTION

The importance of hypoxia as a cause of local failure in radiotherapy was described by the late L.H. Gray and his colleagues in 1953 [1].
It was shown that if a cell was deprived of oxygen its sensitivity to damage by X-rays was reduced by a factor of 2.8. Since many cells in malignant tumours are hypoxic, usually as a result of deficient blood supply, they are relatively radioresistant and survive the doses of radiation that can be delivered without excessive damage to normal, well-oxygenated, adjacent tissue. The first clinical trial of the use of hyperbaric oxygen as a radiosensitizer was reported by Churchill-Davidson and his colleagues in 1955 [2].
One theory of the mode of action of oxygen as a radiosensitizer is that it brings about electron capture, preventing repair of damaged molecules from which electrons have been expelled by irradiation.

Benefit has been reported in some randomized clinical trials of hyperbaric oxygen but not in others. Alternative methods for overcoming the resistance of the hypoxic tumour cell have been explored. These include fractionation, heavy particle therapy and now the use of chemical agents.

The first drug given clinical trial as a radiosensitizer was Synkavit (vitamin K_1) and the results of an apparently favourable clinical trial were reported by Mitchell in 1960 [3].

Radiosensitization has since been reported for many cytotoxic agents, particularly alkylating agents (tretamine), halogenated pyrimidines (5-fluorouracil, bromuridine, floxuridine), actinomycin folic acid antagonists (methotrexate), etc. All these later agents have effects on normally oxygenate cells, both normal and malignant, and so may not give overall benefit in clinical radiotherapy.

More recently, the drug ICRF 159 has been claimed to have a radiosensitizing effect (Hellmann, K., personal communication). It is thought that this may be brought about by promotion of the development of capillary blood vessels in the tumour mass — the so-called angiomesomorphic effect — thus enhancing the local distribution of oxygen.

Since 1963, when Adams and Dewey proposed that a relationship existed between the ability of a few chemical compounds to sensitize hypoxic bacterial cells and the electron affinity of these compounds, much experimental work has been carried out and this was summarized recently by Adams and his colleagues [5].

As a result of these studies two compounds were chosen for clinical evaluation, metronidazole (Flagyl, May & Baker) and Ro 07-0582 (Roche). It is the purpose of this paper to present the results of these preliminary studies.

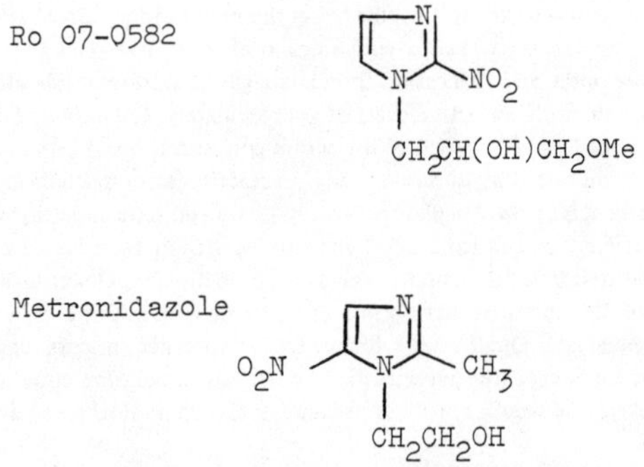

FIG.1. *Structural configurations of Ro 07-0582 and metronidazole.*

MATERIALS

Ro 07-0582 (1-(2-nitro-1-imidazolyl)-3-methoxy-2-propanol) is a 2-nitroimidazole derivative originally developed by Roche in the United States as an anti-trichomonal agent. Metronidazole, a 5-nitroimidazole, is an established agent for the treatment of trichomoniasis and amoebiasis and is under investigation for the treatment of anaerobic infections with organisms such as *Bacteroides fragilis*. The structural configurations of the two substances are shown in Fig. 1. In view of the large doses involved with both compounds, tablets were freshly crushed and suspended in a raspberry juice preparation before administration to make them more palatable.

METHODS

Much discussion has arisen concerning the best method of evaluating a drug which may potentiate the effect of irradiation of tumours in man. In most cases the prolongation of survival time has been taken as the end point. This, although of high clinical relevance — provided that the quality of life in the survival period is maintained or enhanced — is not a measure of the effect of a drug on a particular tumour. It can only give information about groups of patients, properly randomized, with similar tumours. There is never any certainty whether an individual tumour was benefited or not by the treatment given. Further, the cases for clinical trials of new methods of treatment are by necessity advanced. Even if there is a marked improvement in local control, survival will only show, if anything, a slight improvement because of distant disease. In these preliminary studies we had three objectives or questions to answer.

(a) Would the dose of drug that had to be given be tolerated?
(b) Would this dose sensitize hypoxic cells in man?
(c) Could this dose be shown to have any effect in the treatment of tumours in man?

With regard to the first question there was information from records of clinical abuse of the drug that high doses of metronidazole had been taken without ill effect. Metronidazole has also been given in large doses to attempt radiosensitization by Urtasun in Edmonton, Alberta and at Mount Vernon Hospital. A volunteer study with Ro 07-0582 when it was being investigated as a trichomonacide had shown adequate absorption after single oral doses of 250 mg, but animal studies indicated that blood levels of 100–200 μg/ml would be necessary to show a radiosensitizing effect. This meant that doses of the order 6–12 g would be required in man. Consequently, the first step was to administer single oral doses of 1–4 g Ro 07-0582 in healthy human volunteers to determine any clinical effects with these high doses and also to determine the plasma levels and urinary excretion reached after these high doses. If these results were satisfactory then higher doses could be considered for clinical studies.

The second objective, that of determining the response of hypoxic cells in man, was achieved by using areas of skin rendered artificially hypoxic, and by measuring the late pigmentation of such areas with and without Ro 07-0582 after different doses of radiation.

To try to determine any modification of tumour response using Ro 07-0582, studies were made in patients with multiple deposits. In cases where all other methods of treatment had been exhausted it was reasonable to attempt to control disease by irradiating individual deposits by single doses of radiotherapy. Regression and regrowth was carefully followed in nodules treated some with and some without the addition of Ro 07-0582.

RESULTS

1. Volunteer studies

Volunteer studies were undertaken at the Gray Laboratory, Mount Vernon Hospital in 1974. The detailed results have been published by Foster et al. [6]. It is therefore only necessary to summarize them here. Six healthy males took single oral doses of 1, 2 or 4 g Ro 07-0582, in the form of 500-mg tablets, at 10.00 h after a light breakfast more than 2 h previously. Venous blood specimens were taken before dosage and at frequent intervals for the first 3 h, and then at hourly intervals until 18.00 h. Further specimens were taken at 22.00 h and at 10.00 h and 18.00 h on the next day. On the first occasion four subjects each took 1 g. On the second occasion, two weeks later, two subjects took 2 g and two took 4 g. One subject in each of these two groups had previously taken 1 g so that a within-subject comparison could be made between a lower and

TABLE I. SERUM HALF-LIFE OF Ro 07-0582 AFTER SINGLE ORAL DOSES OF 1–4 g

Subject	Age (years)	Wt (kg)	Dose (mg/kg)	Half-life (h)
1	37	70	14.3	9.8
2	37	96.5	10.4	12.8
3	49	76.5	13.0	11.0
4	31	76.5	13.0	10.7
1	37	70	28.6	10.3
5	26	54.5	36.7	11.9
2	37	96.5	41.5	12.5
6	34	79.5	50.3	17.5
				Mean 12.1 ± 2.0

TABLE II. PATIENTS GIVEN 1–10 g Ro 07-0582

Case	Age (years)	Diagnosis	Dose (g)	Dose (mg/kg)	Max. serum level (μg/ml)	Time of peak (h)	Half-life (h)
1	59	Pleomorphic sarcoma Multiple metastases	4.5	81	114	2	9.2
2	33	Carcinoma of cervix Multiple metastases	4.0	96.5	147	2	9.1
3	65	Carcinoma of bladder Multiple metastases	6.5	106	185	1	11.4
4	81	Carcinoma of breast Local skin spread	7.5	114	187	1	17.1
5	58	Carcinoma of breast Multiple metastases	9.0	140	270	2	10.1
6	60	Carcinoma of breast Extension to chest wall	9.25	140	216	1	12.2
7	68	Carcinoma of breast Multiple metastases	7.5	145	159	3	11.8
8	49	Carcinoma of breast Multiple lung metastases	10.0	165	415	2	14.1
9		Soft tissue sarcoma Central necrosis	1.0		23	–	–
10		Soft tissue sarcoma Central necrosis	1.0		29	–	–
							Mean 11.88

higher dose. The concentration of Ro 07-0582 was determined directly by a polarographic technique [7].

Table I gives some details of the age, weight and individual doses of the six subjects together with the serum half-life determined after each dose. It can be seen that the half-life varied from 9.8–17.5 h, one subject having a prolonged serum half-life after the 4-g dose: the mean value was 12.1 ± 2.0 h. The serum half-life was generally independent of dose.

The drug was rapidly absorbed and peak serum levels were usually maintained between 1–4 h after dosing. There was a linear relationship between the drug serum concentration at 4 h and the dose administered expressed in mg/kg. Doses of 10.4–14.3 mg/kg gave serum concentrations of around 20 µg/ml. Doses of 28.6 and 36.7 mg/kg gave figures of around 40 µg/ml. Doses of 41.5 and 50.3 mg/kg gave figures around 60 µg/ml. By extrapolation of a straight-line plot it could be predicted that a dose of 78 mg/kg would be needed to achieve a serum concentration of 100 µg/ml at 4 h, this serum level being that at which radiosensitizing activity was likely to be demonstrated.

There were no significant side-effects reported after the 1-g dose. After the 2-g dose both subjects reported mild insomnia that night. This symptom was more marked in the subjects taking 4 g and it was accompanied by mild gastro-intestinal symptoms. Transient nausea was experienced after the 2- and 4-g doses and the taste of the drug, whilst noticeable, was not particularly unpleasant. There was no obvious interaction with moderate quantities of alcohol taken with meals 3.5 and 10 h after taking the drug.

2. Studies in patients with carcinoma

Ten patients with advanced malignant disease, where the life expectancy was less than twelve months, were given single oral doses of Ro 07-0582 ranging from 1–10 g (22.5–165 mg/kg), see Table II. Two patients received only 1 g and were used for studies of tumour levels of Ro 07-0582. The remaining eight patients were given 4–10 g (81–165 mg/kg), and serum levels in excess of 100 µg/ml were obtained in every case. The maximum mean serum levels were generally in proportion to the dose given with the exception of the patient taking 10 g. The maximum serum level in this patient was unexpectedly high, 415 µg/ml, and it is postulated that doses much above 140 mg/kg may result in flooding of enzyme systems leading to retention and accumulation of Ro 07-0582: doses above 140 mg/kg may thus lead to toxic effects from excess drug.

TABLE III. SIDE-EFFECTS IN PATIENTS GIVEN 4–10 g Ro 07-0582

Case	Dose (g)	Dose (mg/kg)	Bad taste	Nausea/ vomiting	Diarrhoea	Overall tolerance
1	4.5	81	++	0	0	Good
2	4.0	96.5	++	+	0	Good
3	6.5	106	+	++	0	Fair
4	7.5	114	0	+	0	Good
5	9.0	140	+	+	+	Fair
6	9.25	140	++	++	0	Fair
7	7.5	145	++	++	0	Fair
8	10.0	165	+++	+++	+	Poor

We also found that a dose of 140 mg/kg was the upper limit for clinical tolerance from the point of view of gastro-intestinal side-effects (Table III). Some degree of clinical intolerance was seen in all patients and this was largely dose-related. The symptoms could be controlled by the use of anti-emetic drugs although these were only used routinely before treatment in two patients. Our patients, who were all with advanced and metastatic tumour, were probably more likely than the average patient to suffer gastro-intestinal side-effects.

Full blood counts, serum electrolyte and uric acid levels, and standard tests for renal and hepatic function were measured before treatment and at weekly intervals for six weeks. No significant changes were seen.

Particular attention was paid to possible neurological side-effects in view of the reported neurotoxicity in animals given analogous compounds [8]. All patients had full neurological examinations before and after treatment: brain scans and skull X-rays were also performed before treatment in all cases. In one patient an initial brain scan was suggestive of intracerebral lesions but a repeat following treatment was normal: the skull X-ray was normal throughout. Four days after receiving 6.5 g (106 mg/kg) he developed an abnormal Romberg's test. This was probably due to tumour involvement of the right femoral nerve revealed at autopsy. One other patient had mild truncal ataxia before treatment and this improved in the weeks following treatment.

The serum half-life measured from the serum concentrations ranged from 9.1–17.1 h and was consistent with the results of the volunteer studies. One patient had a rather longer half-life than the seven others.

The urinary excretion of the drug was monitored for the first 24 h in seven patients and for 48 h in four. The results are difficult to interpret because of variable fluid intake due to nausea and vomiting in some patients. Our preliminary results indicate that the drug is excreted largely in the urine, partly unchanged and partly as the demethylated metabolite, Ro 05-9963. This latter compound has been synthesized and is a potent radiosensitizer in in-vitro situations.

The results of the tumour level studies are shown in Table IV. The levels achieved, expressed as a per cent of the simultaneous serum concentration, were variable, ranging from 12–92%. The low levels achieved in two of the patients with carcinoma of the breast may have been associated with difficulties encountered in the extraction procedure.

TABLE IV. CONCENTRATION OF Ro 07-0582 IN TUMOUR TISSUE

Case	Tumour	Specimen	Blood contamination	Time after admin.	Dose (mg/kg)	Serum concn (μg/ml)	Tumour concn μg/ml	% of serum concn
5	Carcinoma of breast	Drill biopsy	0	4 h 0 min	9.0	250	36	14
6	Carcinoma of breast	Drill biopsy	0	4 h 5 min	9.25	205	144	70
8	Carcinoma of breast	Drill biopsy	0	4 h 5 min	10.0	296	35	12
9	Soft tissue sarcoma	Aspirate of fluid	+	4 h 0 min	1.0	23	21	92
10	Soft tissue sarcoma	Aspirate of fluid	0	5 h 15 min	1.0	29	18	62

2.1. Radiosensitization of normal and hypoxic skin

The technique used in our patients has been described [9] and the detailed methodology and results achieved have been set out by Dische and his colleagues [10]. Six or eight adjacent areas of skin, 1.5 cm square, are marked out on the skin, usually of the upper anterior aspect of the forearm. Each area was separately irradiated with graded known doses under oxic and hypoxic conditions, without and with the presence of the sensitizer. The sites for the different radiation conditions were randomly allocated to eliminate bias in the later assessments. The radiation source was a strontium-90 plaque supplied by the Radiochemical Centre, Amersham. The active area is 1.5 cm square and the surface dose-rate was 8.37 rad/s with a source uniformity of ± 3.0%. A 0.15% decay correction was applied every six months.

Oxic conditions were ensured by passing oxygen through a plastic bag (with side sleeve for the strontium applicator); the strontium plaque was modified by fitting 1-mm perspex blocks at each inactive corner so that the underlying skin would not be occluded. Hypoxia was achieved by first applying an Esmarch's bandage to the limb starting at the periphery: a sphygmomanometer cuff was then applied and inflated to 200 mgHg before the bandage was removed. Nitrogen was then blown through the plastic bag, instead of oxygen, for 10 min.

Oxic exposures of 800, 900, 1000 and 1100 rad were used whereas 1600 on 2000 rad were delivered under hypoxic conditions. On the following day the procedure was repeated between three and four hours after oral administration of the sensitizer.

Readings of the degree of late skin pigmentation in each area were made by eye by a number of different observers. Each observer independently ranked the areas in order of intensity of pigmentation and the average score for each area was recorded daily. Observations were maintained until the pigmentation faded; this usually occurs after 60–100 days. Six subjects (Nos 1, 2, 3, 5, 7 and 8) were followed. Three other patients had skin-sensitizing tests with metronidazole.

The oxygen enhancement ratio (OER) was calculated as the ratio of the radiation dose given under hypoxia to the determined equivalent oxic dose. The enhancement ratio (ER) was obtained from the ratio of the equivalent oxic dose when the sensitizer was used under hypoxia to the equivalent oxic dose under hypoxia without the sensitizer.

The relative sensitizing efficiency (RSE) was calculated from

$$\text{RSE} = \left(\frac{\text{ER}-1}{\text{OER}-1} \times 100 \right) \%$$

This gives a measure of the efficiency of the sensitizer in restoring the sensitivity of cells protected by hypoxia under the conditions of the study.

TABLE V. SKIN REACTIONS USING Ro 07-0582

Patient	Dose		OER ± SE	ER ± SE	RSE (%)
	g	mg/kg			
1	4.5	80	2.46 ± 0.30	1.55 ± 0.18	38
2	4.0	96.5	1.85 ± 0.26	1.23 ± 0.13	27
3	6.5	106	1.85 ± 0.27	1.25 ± 0.14	30
5	9.0	140	1.95 ± 0.12	1.68 ± 0.10	71
7	7.5	145	1.64 ± 0.23	1.38 ± 0.26	60
8	10.0	165	1.94 ± 0.31	1.50 ± 0.24	53

TABLE VI. SKIN REACTIONS USING METRONIDAZOLE

Patient	Dose		OER	ER	RSE
	g	mg/kg	± SE	± SE	(%)
1	12.0	200	2.22 ± 0.42	1.17 ± 0.16	14
2	11.0	200	2.13 ± 0.15	1.12 ± 0.10	11
3	12.2	200	1.85 ± 0.24	1.13 ± 0.06	15

The skin reactions observed showed both the protection afforded by hypoxia and sensitizatio of hypoxic skin by Ro 07-0582 and, to a lesser extent, metronidazole.

The values obtained for OER, ER and RSE with Ro 07-0582 are shown in Table V. These may be compared with the values for metronidazole in Table VI.

The skin reactions after oxic exposure combined with Ro 07-0582 were not enhanced. The OER's recorded were lower than might have been expected if there was complete hypoxia and it is probable that our technique leaves a low residual concentration of oxygen in the skin. Patients receiving the lower doses of Ro 07-0582 tended to have lower RSE's than the higher dosed group. The RSE values for Ro 07-0582, ranging from 27–71%, were greater than those for metronidazole 11–15%, using doses which were clinically equivalent from the point of view of gastro-intestinal tolerance.

This technique has demonstrated that Ro 07-0582 is a highly effective sensitizer of normal cells rendered artificially hypoxic.

It is clearly demonstrated that Ro 07-0582 will sensitize hypoxic cells in man but we must be careful in extrapolating these findings to the situation in human tumours. Different tumours are likely to have different populations of hypoxic cells and there may be differences of diffusion into these cells compared with the conditions of this study on normal skin.

2.2. Radiosensitization of malignant tumours

To measure a quantitative response we attempted to assess the delay imposed on the regrowth of tumours where local eradication had not occurred [11, 12]. Seven of our patients needed palliative treatment for multiple metastases and had measurable lesions which were suitable for treatment with single doses of radiation; these seven were selected for serial measurement of tumour regression and regrowth; one group of metastases in each patient was treated with Ro 07-0582 + radiation and the remainder were treated with radiation alone.

Three patients had diffuse infiltration of skin and, in each, a number of areas of skin of equal size were treated with different doses. One patient died before any differences were detectable. In one patient, the tumour appeared to recur later in the areas treated with Ro 07-0582 and radiat than in areas receiving the same dose of radiation but no drug. Transient regression was observed i all areas in the third patient who died six weeks after treatment.

Three patients had multiple subcutaneous nodules. One died before any valid assessment cou be made. One, with deposits of pleomorphic sarcoma, showed increased regression in the four nod treated with 800 rad + 81 mg/kg Ro 07-0582 compared with three nodules treated with 960 rad a Regrowth started after approximately 130 days but the patient died before definite conclusions co be drawn.

The third patient in this group had 21 measurable skin nodules from carcinoma of the cervix uteri which had been surgically removed 5 years previously. Seven nodules (mean diameter 14.28 ± 2.25 mm) were treated with 1120 rad: seven nodules (mean diameter 11.47 ± 1.15 mm) were treated with 960 rad: seven nodules (mean diameter 11.99 ± 0.53 mm) were treated with 800 rad between three and four hours after taking 96.5 mg/kg Ro 07-0582 when the serum concentration was about 147 μg/ml. The volume of the tumours in all three groups diminished in a similar manner until day 21 after which regrowth became evident. In the regrowth period the volumes of the groups of tumours treated with 800 rad + Ro 07-0582 increased at the same rate as those treated with 960 rad alone. There was a slower increase in volume in the group treated with 1120 rad. This suggests that Ro 07-0582 enhanced the response to radiation by a factor of 20% in this patient (ER = 1.2).

The seventh patient had bilateral pulmonary metastases from a locally recurring carcinoma of the breast. A dose of 900 rad was applied to the left lung field and the same dose was applied to the right lung field three hours after administration of 165 mg/kg Ro 07-0582. Repeated measurements were made of the six discrete metastases (mean diameter 18.7 ± 3.9 mm) on the right side and the thirteen metastases (mean diameter 17.2 ± 0.9 mm) on the left. No enhancement of effect was seen from the administration of the drug and indeed regrowth was slightly faster on this side. Fuller details of this work will be published shortly [13].

DISCUSSION

The studies described have shown that single oral doses of up to 140 mg/kg Ro 07-0582 are fairly well tolerated. An enhancement ratio of up to 1.68 ± 0.1 has been demonstrated in normal skin cells rendered artificially hypoxic. Among seven patients it was possible to show some qualitative enhancement of radiation effect upon tumour in two and a measurable enhancement ratio of 1.2 in one patient using a dose of 96.5 mg/kg. The significance of this result is that it is the first demonstration of a quantifiable radiosensitizing effect by a drug in a human tumour.

Despite these promising results we have yet to establish the value of Ro 07-0582 in clinical radiotherapy. We have yet to ascertain the correct dosage, its value with courses of radiotherapy and to learn of the toxicity likely to be encountered if repeated doses are given.

Fractionation of radiation doses is normal practice in radiotherapy to increase efficacy in eradication of tumour without increasing damage to normal tissues. To follow this would require giving repeated high doses of a chemical radiosensitizer and it is possible that this may lead to unacceptable toxicity. The neurotoxicity of the nitroimidazoles to animals has been documented by Schärer [8] and it will be necessary to define very carefully the maximum tolerated repeated doses in man. Similarly, it is very important that a wide range of tumours be studied in detail so that the possibilities of this new approach can be defined with more precision. In the small series reported here there was an increased effect of only + 20% in one patient whilst in another there was no measurable effect. In the latter patient there may have been good reasons for this, such as poor entry of the drug to the tumour areas because of previous irradiation, but the lack of effect needs examination.

Again, it has been shown [14] that nitroimidazoles may have a mutagenic action in bacteria. The relevance of this to clinical practice — particularly to those already suffering from neoplastic disease — must be assessed.

We plan to undertake studies using fractionated radiation with and without repeated doses of Ro 07-0582. No wide extension of trials will be justified until we have data from such studies.

We have already shown that, in normal skin, an ER of 1.7 is possible. This is the same as that achieved by neutron beam therapy. Such treatment requires expensive apparatus and at this time it is difficult to apply to all sites. The promise of the chemical agents is that they may be used in the treatment of all patients in all radiotherapy centres without further investment in apparatus.

ACKNOWLEDGEMENTS

Many people have been involved in the production of the results that we have obtained. Dr. J.F. Fowler, Dr. G.E. Adams, Dr. J.L. Foster and Dr. I.R. Flockhart of the Gray Laboratory of the Cancer Research Campaign gave invaluable scientific assistance. Dr. A. Gray assisted in the clinical studies. Dr. G.D. Zanelli assisted with both the clinical and physical estimations. Dr. R.H. Thomlinson was assisted by Miss L.M. Errington in the measurement of tumour nodules. Dr. C.E. Smithen and Dr. W.M. Parkes gave chemical and toxicological advice. Mrs. Irene Lansley gave statistical help. Without the help of Sister M. Lee, Miss S. Warwick and the staff of radiographers and nurses and our colleagues in the Departments of Pathology and Physics this paper would not have been possible.

REFERENCES

[1] GRAY, L.H., et al., Br. J. Radiol. **26** (1953) 638.
[2] CHURCHILL-DAVIDSON, I., et al., Lancet **1** (1955) 1091.
[3] MITCHELL, J.S., Studies in Radiotherapeutics, Oxford (1960) 87 and 160.
[4] CATTERALL, M., et al., A randomised clinical trial of fast neutrons on X or γ-rays in the treatment of advanced tumours of the head and neck, Br. Med. J. **2** (1975) 653.
[5] ADAMS, G.E., DENEKAMP, J., FOWLER, J.F., "Biological basis of radiosensitization by hypoxic-cell radiosensitizers", Proc. 9th Int. Congr. Chemotherapy, London, July 1975 (in press).
[6] FOSTER, J.L., et al., Serum concentration measurements in man of the radiosensitizer Ro 07-0582: some preliminary results, Br. J. Cancer **31** (1975) 679.
[7] KANE, P.O., Polarographic methods for the determination of two anti-protozoal nitro-imidazole derivatives in materials of biological origin, J. Polarogr. Soc. **7** (1961) 758.
[8] SCHÄRER, K., Selective alterations of Purkinje cells in the dog after oral administration of high doses of nitroimidazole derivatives (in German), Verh. Dtsch. Ges. Pathol. **56** (1972) 407.
[9] DISCHE, S., ZANELLI, G.D., Skin reaction — a quantitative system for measurement of radiosensitization in man, Clin. Radiol. (in press).
[10] DISCHE, S., GRAY, A.J., ZANELLI, G.D., Clinical testing of the radiosensitizer Ro 07-0582. II. Radiosensitization of normal and hypoxic skin, Clin. Radiol. (in press).
[11] THOMLINSON, R.H., Br. J. Cancer **14** (1960) 555.
[12] THOMLINSON, R.H., CRADDOCK, E.A., Br. J. Cancer **21** (1967) 108.
[13] THOMLINSON, R.H., et al., Clinical testing of the radiosensitizer Ro 07-0582. III. Response of tumours, Clin. Radiol. (accepted for publication).
[14] VOOGD, C.E., et al., The mutagenic action of nitroimidazoles. II. Effects of 2-nitroimidazoles, Mutat. Res. (1975) 149.

THE PROBLEMS OF RADIOSENSITIZATION

E.F. ROMANTSEV, A.V. NIKOLSKIJ
Institute of Biophysics,
Ministry of Health of the USSR,
Moscow, USSR

Short communication

One of the main problems of modern radiobiology is associated with the use of chemical compounds (radiosensitizers) for sensitizing biological material to the action of ionizing radiation. The significance of these compounds is determined by their long-term use for radiotherapy of malignant diseases. At present, some of these compounds such as halogenated derivatives of pyrimidine bases, actinomycin D, vicasol and others, have been effectively applied in radiotherapy of various tumour types. The search for new radiosensitizers and the study of their action mechanism are of significant interest for theoretical radiobiology as well, since this knowledge opens the way for understanding primary challenging biochemical mechanisms of ionizing radiation effect on the living cell.

Radiation protection and radiosensitization may be regarded as two aspects of the same problem of radiosensitivity. But, whereas in radioprotection we are always interested in the increase of radioresistance of the body as a whole, the use of radiosensitizers is limited to the increase of lethal effect of ionizing radiation on tumour cells **only**. Considering the modern techniques of local focused irradiation, this problem is not very difficult. However, the task of selective effect has not been solved yet for most compounds. Promising possibilities in this connection are associated with liposomal complexes whose application will probably allow us to increase the concentration of radiosensitizer in tumour tissue.

Attemps to attribute the radiosensitizing effect of different chemical substances to a single-action mechanism have failed. From the point of view of the "sulphydryl" hypothesis, the increase of radiation effect is due to the ability of radiosensitizers to decrease directly or indirectly the concentration of intracellular thiols (which are natural radioprotectors) [1,2]. There is some evidence against this concept, but as a result of persistent study of the compounds inhibiting SH-groups such promising radiosensitizers as azoether (methylphenyldiazenecarboxylate) [3], diamide (bis-(N, N-dimethylamide)diazenedicarboxylic acid) [4], and cyclohexene derivatives [5] have been found. In our experiments a representative of the cyclohexene group gave an almost two-fold decrease of non-protein thiol concentration in ascites lymphosarcoma NK/Ly of mice, and increased tumour radiosensitivity by 20–30% (see Tables I and II). The correlation of radio-sensitization with electron-acceptor abilities of chemical compounds [6] also allowed us to find a number of new agents which modify radiation effects [7–9].

In our opinion the most complete study of radiosensitivity modifications requires a careful investigation of the biochemical changes which occur under the influence of radiomodifiers as a result of essential rearrangement in complex and co-ordinated biochemical processes induced by them up to the moment of irradiation. The study of these changes is very important and we believe that the hypothesis suggested by Langendorff [10] and developed by others [11,12] is worth noting in this connection. According to this hypothesis, cell radiosensitivity depends on the content of adenosine 3', 5'-cyclic monophosphate (cyclic AMP) and thus on the functioning of the enzymes involved in the metabolism of this compound, i.e. adenyl-cyclase and phosphodiesterase. The first experimental evidence in favour of this suggestion has already been obtained.

TABLE I. INFLUENCE OF CYCLOHEXENE ON LEVEL OF NON-PROTEIN THIOLS IN CELLS OF ASCITES LYMPHOSARCOMA NK/Ly OF MICE AS A PER CENT OF THE CONTROL (100 ± 12.9)

Time after injection of cyclohexene (min)	Level of non-protein thiols as a per cent of the control
15	68.6 ± 8.2
30	49.2 ± 5.5
60	70 ± 1.6

Note: Cyclohexene (100 mg/kg body weight) injected i.p. on 4th day after transplantation of tumour. Non-protein thiols determined according to method described in Ref. [16].

TABLE II. INFLUENCE OF CYCLOHEXENE AND GAMMA IRRADIATION ON AMOUNT OF TUMOUR CELLS IN ASCITES LYMPHOSARCOMA NK/Ly

Experimental conditions	Amount of the tumour cells $\times 10^9$
Control	1.46 ± 0.12
Cyclohexene, 100 mg/kg i.p.	1.44 ± 0.15
Irradiation, 600 R	0.87 ± 0.04
Irradiation 600 R + cyclohexene	0.59 ± 0.03

Note: On 4th day after tumour transplantation the mice were irradiated with ^{60}Co, 346 R/min. The amount of ascites cells was calculated on the 8th day after tumour transplantation.

For example, it is known that imidazole, which activates phosphodiesterase and reduces the level of cyclic AMP, is a radiosensitizer [13], whereas theophylline, which inhibits this enzyme, is a radioprotector [14]. Unfortunately, we do not yet know which compounds have a specific influence on adenyl-cyclase activity. Theophylline inhibits it by 20–30%, but its inhibiting effect on phosphodiesterase is much greater resulting in an increase in the cyclic AMP level [15]. Therefore, we believe that compounds actively involved in cyclic AMP metabolism will be of much value in the search for radiomodifiers.

REFERENCES

[1] BRIDGES, B.A., Advances in Radiation Biology, 3rd Edn, Academic Press, New York (1967).
[2] TARASENKO, A.G., NEKRASOVA, I.V., GRAEVSKIJ, E.Ya., Radiobiologiya (in Russian) **10** 2 (1970) 198.
[3] HARRIS, J.W., PAINTER, R.B., Int. J. Radiat. Biol. **15** 3 (1969) 289.
[4] HARRIS, J.W., POWER, J.A., Radiat. Res. **56** 1 (1973) 97.
[5] JAMES, S.P., JEFFERY, D.J., WHITE D.A., Biochem. Pharmacol. **20** (1971) 897.
[6] ADAMS, G.E., Adv. Radiat. Chem. **3** (1969) 125.
[7] FOSTER, J.L., WILSON, R.L., Br. J. Radiat. Biol. **46** (1973) 234.
[8] CHAPMAN, J.D., REUVERS, A.P., BORSA, J., ibid. **46** (1973) 623.
[9] ASQUITH, J.C., WATTS, M.E., PATEL, K., SMITHEN, C.E., ADAMS, G.E., Radiat. Res. **60** (1974) 108.
[10] LANGENDORFF, H., Strahlentherapie **140** 4 (1970) 428.
[11] PRASAD, K.N., Int. J. Radiat. Biol. **22** 2 (1972) 187.
[12] MITZNEGG, P., Int. J. Radiat. Biol. **24** 4 (1973) 339.
[13] ZHIVOTOVA, N.I., FILIPPOVICH, I.V., ROMANTSEV, E.F., VOPROSIJ, Med. Khim. (in Russian) 22 2 (1976).
[14] PAZDERNIK T.L., UYEKI E.M., Int. J. Radiat. Biol. **26** 4 (1974) 331.
[15] POHL, S.L., BIRNBAUMER, L., RODBELL, M., Science **164** 3879 (1969) 566.
[16] ELLMAN, G.L., Arch. Biochem. Biophys. **82** (1959) 70.

CHANGES IN THE RATIO OF ACTIVITY OF REPARATIVE AND REPLICATIVE ENZYMES OF DNA SYNTHESIS AS A BASIS FOR THE SEARCH FOR RADIOPROTECTIVE DRUGS

I.V. FILIPPOVICH, E.F. ROMANTSEV
Institute of Biophysics,
Ministry of Health of the USSR,
Moscow, USSR

Short communication

The concept of biological multiplication of initial radiation injury has gained wide acceptance. This concept is based on the hypothesis that injury of unique or control systems of cells is reflected in the functioning of the systems liable to control [1] and is probably of value for such unique structures as DNA molecules. The repair of DNA (the control system) must result in less injury of the systems liable to control. So, any change in the intensity of the repair processes by means of various actions may result in a favourable effect.

It is known that irradiation leads to structural changes in DNA which can be repaired partially. For normal functioning of the cell, DNA replication has to take place on a pre-existing template. If the rate of the repair processes exceeds both the rate of replication and the intensity of radiation injury of the template, the cell will survive [2]. In this case, the effects resulting in a decrease of initial damage to the DNA as well as of the rate of replication must lead to an increase in survival rate.

We performed appropriate experiments on the effect of specific inhibitors of replicative DNA polymerase, such as nalidixic acid[1] and ara-C[2] [3,4] on the survival rate of irradiated thymocytes (see Table I).

TABLE I. EFFECT OF PRE-TREATMENT OF THYMOCYTES WITH NALIDIXIC ACID AND ARA-C ON SURVIVAL OF IRRADIATED CELLS

Experimental conditions	Amount of dead cells:	
	in ml of suspension	in per cent
Control (non-irradiated) (10)	$9.75 \times 10^7 \pm 1.15 \times 10^7$	11.1 ± 1.0
Irradiation (800 R) (10)	$2.45 \times 10^8 \pm 2.05 \times 10^7$	28.6 ± 1.5
Pre-incubation with nalidixic acid and irradiation (5)	$1.27 \times 10^8 \pm 2.75 \times 10^7$	15.2 ± 1.2
Pre-incubation with ara-C and irradiation (5)	$1.02 \times 10^8 \pm 2.10 \times 10^7$	12.4 ± 0.6

The numbers in parentheses refer to the number of experiments.

---

[1] Nalidixic acid: 1-ethyl-1, 4-dihydro-7-methyl-4-oxo-1, 8-naphthyridine-3-carboxylic acid.
[2] Ara-C: 1-β-D-arabinofuranosylcytosine.

As seen from the Table, pre-incubation of cells with specific inhibitors of replicative DNA polymerase protects the cells against radiation. Therefore, temporary inhibition of DNA replication contributes to the repair of a damaged template because of the decrease in the number of mistakes during replication. Up to the moment when inhibition of DNA replication disappears, most of the template damage is repaired and DNA continues to function as a template.

REFERENCES

[1] KUZIN, A.M., Structure-metabolic hypothesis in radiobiology (in Russian), Science (1970).
[2] BROWN, P.E., Nature (London) 213 (1967) 363.
[3] SCHNECK, P.K., STANDENBAUER, W.L., HOFSCHNEIDER, P.H., Eur. J. Biochem. 38 (1973) 130.
[4] SCHRECKER, A.W., SMITH, R.G., GALLO, R.C., Cancer Res. 34 (1974) 286.

CONCLUSIONS AND RECOMMENDATIONS

RADIATION SENSITIZING AGENTS

Within the last decade, various classes of chemical agents have been found which increase the efficiency of radiation-induced cellular damage. This is important in two respects:

(i) Exposure to low doses of radiation occurs in routine diagnostic radiology, clinical investigations with radioactive isotopes and in the industrial applications of nuclear energy. It is likely that some chemicals and drugs can act as sensitizers at these low radiation levels and can therefore constitute a health hazard.
(ii) At present, roughly half of all cancer patients receive radiotherapy and many are successfully treated. However, failures in local control do occur and even a modest improvement in the relative radiosensitivity of tumours would result in a substantial improvement in control of local disease. It is for this reason that there is increasing interest in the development and application of radiosensitizers in this field.

This Advisory Group emphasizes the distinction drawn by the previous panel on *Modification of Radiosensitivity in Biological Systems with Particular Emphasis on Chemical Radiation Sensitization and its Use in Radiotherapy* in collaboration with the WHO (Stockholm, 25–29 June 1973) between true radiosensitizing agents and those drugs which are used solely as adjunctive chemotherapeutic agents.

A drug suitable for use as a radiosensitizer in radiotherapy must result in improved tumour control. This should be achieved with:

(a) as little toxicity as possible;
(b) ease of administration and maintenance of effective drug levels;
(c) as great a tumour versus normal tissue selectivity as possible; either by selective accumulation in tumours or by a selective effect on tumour cells;
(d) a high degree of predictability of effectiveness and tolerance for different tumours and hosts;

and, preferably without:

(a) adverse interaction with other adjuvant therapy;
(b) adverse effect on the immune system.

Compounds not intended for clinical use, but developed for studies of basic mechanisms, need not fulfil all of these criteria. However, for such studies, preference should be given to situations where unambiguous interpretation of the results is possible.

PRESENT STATUS AND RESEARCH PROGRESS

Classification of radiosensitizers

At the previous panel meeting in Stockholm, radiation sensitizers were classified into six groups. It was accepted by the panel that the classification did not exclude the existence of other types of drug, nor did it imply that the groups were mutually exclusive.

The present Advisory Group decided that, in view of the many developments in the general field of radiosensitizers, the method of classification should be amended as follows:

CONCLUSIONS and RECOMMENDATIONS

(1) Radiosensitizers specific for hypoxic cells
   (a) Electron-affinic agents
   (b) Membrane-specific agents

(2) Analogues of DNA precursors
   (a) Incorporated into DNA
   (b) Not incorporated into DNA

(3) Radiation-activated cytotoxic compounds

(4) Factors which modify cellular regulatory processes
   (a) Inhibitors of repair
   (b) DNA-binding and intercalating compounds
   (c) Inhibitors of natural radioprotection
   (d) Hyperthermia

(1) Radiosensitizers specific for hypoxic cells

(a) Electron-affinic agents

Interest in drugs which specifically radiosensitize hypoxic cells rests on the proposal that (i) hypoxic cells are present in a significant proportion of human tumours and (ii) that they are a major obstacle to improvement in the local control of cancer treated with radiotherapy.

Methods proposed to overcome the relative radioresistance of hypoxic cells in tumours include the applications of different fractionation regimes, radiotherapy with neutrons or with other high energy particles, and radiotherapy in normobaric and hyperbaric oxygen. All three methods have been examined, or are in the process of being examined, in randomised controlled clinical trials.

The fourth approach to the hypoxia problem is to use sensitizing drugs which are effective against hypoxic cells in tumours without increasing the radiation sensitivity of well-oxygenated normal tissue.

Considerable research progress has been made in this field during the period since the last panel meeting in 1973. Many different types of chemical structure have been investigated including nitrofurans, nitroimidazoles and other nitroheterocyclic compounds, and have been shown to be specific hypoxic cell sensitizers in vitro. Basic studies showing a quantitative relationship between one-electron reduction potential (electron affinity) and sensitizing efficiency have helped in the search for new drugs.

At the present time, the nitroimidazoles appear to have the best prospects of this class since they satisfy, to a large extent, the criteria which should be met before a compound can be investigated clinically. These drugs are tolerated at high doses in man; they have suitable pharmacokinetic properties and they are widely distributed in tissues. Large sensitization effects have been demonstrated in a variety of solid mouse tumours containing hypoxic cells.

The pharmacological and toxicological properties of the 2- and 5-nitroimidazoles have been widely investigated in several species including man. Pilot clinical studies with compounds of this type are in progress.

(b) Membrane-specific agents

Recently, some drugs in the general class of anaesthetics, analgesics and tranquillizers, all showing a common property of specific cell-membrane interaction, have been shown to sensitize hypoxic bacteria and mammalian cells to ionizing radiation. A promising observation is that some of these drugs, when in combination, have a sensitizing effect on bacteria greater than that of oxygen. The mechanism of action of these compounds is not yet understood and much basic research is required both under in-vitro and in-vivo conditions. This will provide guidelines for the screening of such compounds.

A possible advantage of this group of drugs which could facilitate any future clinical applications is that their toxicology and pharmacology are well known.

(2) Analogues of DNA precursors

(a) Analogues of DNA precursors, such as bromodeoxyuridine (BUdR), are incorporated into DNA and lead to cellular sensitization by increasing radiation damage in the DNA. With regard to clinical applicability of this group of substances, there appears to be little change in the position as stated by the previous panel.

(b) Other analogues of DNA precursors may act by inhibition of nucleotide synthesis and thus indirectly reduce the efficiency of radiation-induced DNA repair processes.

(3) Radiation-activated cytotoxic compounds

With respect to the position stated by the previous panel, no new facts have been reported on the fundamental aspects of radiosensitization by iodine-containing compounds.

With regard to water-soluble and non-toxic iodinated radiological contrast media (e.g. iothalamic acid), further results have been obtained on their radiosensitizing activity in single-cell systems. Promising observations have also been reported indicating the influence of iothalamic acid on the radiation control of solid mouse tumours. These studies require follow-up in many systems and should include low radiation doses and dose-rates.

(4) Factors that modify cellular regulatory processes

Agents that sensitize by modifying endogenous protective or repair mechanisms provide important clues to the mechanisms of radiation resistance and radiosensitization. However, the knowledge of these basic mechanisms is inadequate. One of the best examples of sensitization by interference with endogenous regulatory processes is to be found in the relationship between DNA repair and cell survival. This relationship is far better understood in bacteria than in mammalian cells. It is not known which of the several types of DNA repair could be the most sensitive target in mammalian cells, but it is apparent that agents which bind to DNA or which interfere with DNA repair, are effective sensitizers. Introduction into the cell of exogenous enzymes (e.g. mycoplasma endonuclease) may lead to sensitization by inhibiting unscheduled DNA synthesis.

Radiosensitization may also result from inhibition of endogenous hydrogen donors (e.g. low molecular weight thiols and reduced pyrimidine nucleotides); inactivation of sulfhydryl-dependent repair enzymes; or interference with various other components of endogenous repair functions (e.g. cAMP). Further studies with the radiosensitizers diamide and N-ethyl maleimide, suggest that interference with such sulfhydryl-dependent functions can result in radiosensitization. There are other endogenous systems which, if modified, could also result in radiosensitization. These include the energy-producing and the melanin systems. Recent research on the latter suggests that radioresistance can be substantially decreased in a variety of pigmented tissues including melanotic tumour cells in vitro and in vivo. These investigations have led to a pilot clinical study in patients with melanoma.

Further work is required on other factors including those which regulate the cell cycle, structural components of the cell, and on agents or drugs which concentrate in tumours. Some treatment modalities may act by more than one mechanism as, for example, hyperthermia.

CONCLUSIONS and RECOMMENDATIONS

RADIATION PROTECTING AGENTS

It has been firmly established during the last quarter of a century that harmful effects of ionizing radiation in biological systems, including mammals, can be diminished either by pre-treatment with certain chemicals or by selective modification of endogenous protective substances. With respect to the former, however, the intensive research has failed to produce radioprotective drugs which can be satisfactorily used in man.

Radioprotective substances could be useful for:

(1) **Healthy people** who have to be exposed to large doses of more or less homogeneous total-body irradiation, for instance during emergency operations;
(2) **Patients** exposed to large doses of local irradiation in the course of tumour treatment, or moderate doses of irradiation due to diagnostic applications of radiation sources.

The following criteria will have to be met before any radioprotector can be considered for human use:

(a) it should be effective in doses well below the amounts producing toxic side effects;
(b) it should be effective at radiation dose-levels which might lead to severe early and late somatic consequences or even genetic effects;
(c) its effectiveness should last for at least one hour or more;
(d) in addition, those to be used in the course of radiotherapy should have a differential effect on tumour versus normal tissues either by selective mechanism of action, or selective accumulation, or by a selective mode of administration.

PRESENT STATUS

The great majority of well-established radioprotective substances that proved to be the most effective in animal experiments, e.g. cysteamine (MEA), cystamine, aminoethylisothiuronium (AET), serotonin, etc. are too toxic for human beings and this seriously limits their applicability. Nevertheless, a few compounds have been found recently, such as some phosphothioates and derivatives of mercaptopropylglycine (MPG), that have relatively low toxicity. Appropriate combinations of radioprotective substances having different mechanisms of action, also appear suitable for reduction of toxicity and enhancement of protective effectiveness.

Radioprotective substances effective in animals may offer considerable protection against the symptoms of radiation injury and decrease the rate of haematological and gastrointestinal death. Significant protection can also be obtained against late consequences of irradiation, such as life-shortening, induction of leukaemia and cancer, nephrosclerosis and other late effects. There is evidence that some of the teratogenic and genetic effects of radiation can also be diminished by protective substances.

The majority of radioprotectors presently available provide a protection of only limited duration, generally less than one hour. However, some of the new compounds, e.g. MPG, appear to be active for a longer period. It should be mentioned that a direct correlation has been found between the concentration of radioprotective substances in radiosensitive tissues and their protective effects. This suggests the idea that a protective level in the tissues might be maintained by an appropriate pharmaceutical formulation of these compounds which would assure a prolonged period for absorption and long-lasting activity.

Several aminothiols have been reported to accumulate preferentially in tissues with normal blood supply, at least in the early period after administration. This might lead to a higher protection of normal surrounding tissues than of the usually poorly vascularized solid tumours.

In-vitro and in-vivo experiments have shown that some of the protective substances can influence radiosensitivity of normally oxygenated cells more than that of hypoxic or anoxic cells. If this can be confirmed in intact animals, a selective protection of normal tissues versus tumour tissues might be expected.

Local or topical application of protective substances as well as intra-arterial introduction of such chemicals can offer other possibilities of protecting selected areas or tissues.

Protection of the immune system by chemical means has already been demonstrated in irradiated animals. If the immune system plays an important role in controlling induction and proliferation of tumour cells, protection of this system might be a powerful adjuvant to radiation therapy.

RESEARCH PRIORITIES

The need to obtain less toxic and more efficient protective drugs either among the derivatives of radioprotectors presently available, or among new classes of chemicals, remains a major objective of research. Systematic analysis of correlations between the chemical structure and biological activity of radioprotective compounds is to be preferred to random screening of various chemicals. In addition to this, further research on combinations of radioprotectors having different mechanisms of action is advisable.

Comparative studies on normal tissues and tumours are urgently needed to verify the selective action of radioprotectors that would facilitate their exploitation in tumour therapy.

More information is also necessary on the protection of the immune system of irradiated animals.

Dependence of the efficiency of radioprotectors on the quality of radiation and mode of irradiation are subject areas where further research efforts would certainly be justified.

Most of our knowledge on the mechanism of radioprotection (and radiosensitization) has stemmed from research on the damage and repair of biological systems from the effects of all types of radiations. Therefore, continued fundamental research on these basic mechanisms is highly recommended.

RECOMMENDATIONS TO THE IAEA

(1) The Advisory Group recommends that further research towards the development and identification of effective radiosensitizers and radioprotectors should be encouraged, co-ordinated and supported.

(2) The activities and progress in the different lines of research outlined here have been unequal. In view of this, it should be emphasized that the apparently more rapid development of clinically applicable substances in one field should not justify a decreased attention to the other fields. However, it might be considered whether this inequality would not justify a change in the breadth of the field to be discussed by each future panel of experts in this field. It would seem preferable to have occasional advisory groups on more limited topics with adequate scientific and clinical representation for that subject.

(3) The Advisory Group expresses the desirability of contact being established by the IAEA with WHO or the IUCC or both for possible joint implementation of these recommendations wherever appropriate.

CONCLUSIONS and RECOMMENDATIONS

(4) The Advisory Group feels that a bibliography containing references assembled from the world literature of the last 25 years of research on chemical radioprotectors (and radiosensitizers separately) would be a very useful contribution to any further development in this field. The usefulness of such a publication is indicated by the fact that similar bibliographies on other subjects, for instance "Neutrons in radiation biology and therapy", are being profitably used by interested scientists.

LIST OF PARTICIPANTS

ADAMS, G.E.   Molecular Radiobiology,
CRC Gray Laboratory,
Mount Vernon Hospital,
Northwood, Middlesex HA6 2RN,
United Kingdom

ALTMANN, H.   Österreichische Studiengesellschaft
    für Atomenergie G.m.b.H.,
Lenaugasse 10,
1082 Vienna,
Austria

BRUSTAD, T.   Norsk Hydro's Institute for Cancer Research,
The Norwegian Radium Hospital,
Montebello, Oslo 3,
Norway

DJORDJEVIĆ, O.   Department of Radiobiology,
Institute of Nuclear Sciences "Boris Kidrič,"
P.O. Box 522,
11001 Belgrade,
Yugoslavia

DUNCAN, J.J. KOFI   College of Medicine,
Lagos University Teaching Hospital,
P.M. Bag 12003, Lagos,
Nigeria

FRITZ-NIGGLI, Hedi   Strahlenbiologisches Institut der
    Universität Zürich,
August Forel-Strasse 7,
8008 Zürich,
Switzerland

HARRIS, J.W.   School of Medicine,
Laboratory of Radiobiology,
University of California,
San Francisco, CA 94143,
United States of America

LIST OF PARTICIPANTS

HUG, O.	Strahlenbiologisches Institut,
	Universität München,
	Bavariaring 19,
	8 Munich 2,
	Federal Republic of Germany

ŁUKIEWICZ, S.	Department of Biophysics,
	Institute of Molecular Biology,
	Jagiellonian University,
	31-001 Cracow,
	Poland

MAISIN, J.R.	Radiobiology Department,
	C.E.N./S.C.K.,
	2400 Mol,
	Belgium

PUTTEN, L.M., van	Radiobiologisch Instituut TNO,
	Lange Kleiweg 151,
	Rijswijk, Z.H.,
	The Netherlands

QUINTILIANI, M.	Biochemistry Department,
	Istituto Superiore di Sanità,
	299, Viale Regina Elena,
	00161 Rome,
	Italy

RÉVÉSZ, L.	Radiobiology Unit,
	Department of Tumor Biology,
	Karolinska Institute Medical School,
	104 01 Stockholm 60,
	Sweden

RIKLIS, E.	Radiobiology Department,
	Nuclear Research Center — Negev,
	P.O. Box 9001 — Beer Sheva,
	Israel

ROMANTSEV, E.F.	Institute of Biophysics,
	Ministry of Health of the USSR,
	Moscow,
	USSR

SUGAHARA, T.	Department of Experimental Radiology,
	Faculty of Medicine,
	Kyoto University,
	Kyoto,
	Japan

LIST OF PARTICIPANTS

SUNDARAM, K.   Bio-medical Group,
Bhabha Atomic Research Centre,
Trombay, Bombay 400 085,
India

SZTANYIK, L.B.   "Frédéric Joliot-Curie"
National Research Institute
for Radiobiology and Radiohygiene,
Pentz K.u. 5,
Budapest 22,
Hungary

REPRESENTATIVE OF ROCHE PRODUCTS LTD

LENOX-SMITH, I.   Clinical Research Dept.,
Roche Products Ltd.,
P.O. Box 8,
Welwyn Garden City,
Hertfordshire AL7 3AY,
United Kingdom

OBSERVERS

GETOFF, N.   Institut für Theoretische Chemie und
Strahlenchemie der Universität Wien,
Währinger Strasse 38,
1090 Vienna,
Austria

KOSTIĆ, L.   Department of Radiobiology,
Institute of Nuclear Sciences "Boris Kidrič,"
P.O. Box 522,
11001 Belgrade,
Yugoslavia

LOCKER, A.   Institut für Theoretische Physik
der Technischen Hochschule in Wien,
Karlsplatz 13,
1040 Vienna,
Austria

YEVDAKOW, V.P.   Institute of Biophysics,
Ministry of Health of the USSR,
Moscow,
USSR

LIST OF PARTICIPANTS

CONSULTANTS

BLEEHEN, N.M.
The Medical School,
Hills Road,
Cambridge CB2 2QH,
United Kingdom

COPPEY, J.
Institut du Radium,
Section Biologie,
Rue d'Ulm,
75005 Paris,
France

LOHMANN, W.
Institut für Biophysik der
    Justus-Liebig Universität,
Strahlenzentrum,
Leihgesternerweg 217,
63 Giessen,
Federal Republic of Germany

SCIENTIFIC SECRETARY

SINGH, B.B.
Section of Radiation Biology,
Division of Life Sciences,
International Atomic Energy Agency,
Kärntner Ring 11,
1011 Vienna,
Austria